SUPERCHARGE ME

Also by Eric Lonergan and published by Agenda
Angrynomics (with Mark Blyth)

SUPERCHARGE ME

Net Zero Faster

Eric Lonergan and Corinne Sawers

agenda
publishing

First published in 2022 by Agenda Publishing

Agenda Publishing Limited
The Core
Bath Lane
Newcastle Helix
Newcastle upon Tyne
NE4 5TF

www.agendapub.com

ISBN 978-1-78821-519-0 (paperback)
ISBN 978-1-78821-520-6 (ePDF)
ISBN 978-1-78821-521-3 (ePUB)

British Library Cataloguing-in-Publication Data
A catalogue record for this book is
available from the British Library

Typeset in Nocturne by Patty Rennie

Printed and bound in the UK by TJ Books

Contents

Acknowledgements

Many people have contributed in different ways to this book. At every stage we have received rapid, insightful, and direct feedback from Shelley and John Sawers. A huge thank you. We have been extremely fortunate to have had detailed comments on all, or parts, of the text from Soumaya Keynes, James Mackintosh, John O'Sullivan, Mark Blyth, Rupert Taylor, Marc Beckenstrater, and Sinead O'Sullivan. We cannot thank you enough. We both learnt a great deal from Alex Hall. His feedback, research, and thoughtful perspectives have been invaluable. Our thinking has also been heavily influenced by a number of conversations with Adair Turner. His clarity of thought in this field is unrivalled.

Discussions on climate change and policy with many friends and colleagues over the years have informed our thinking in ways which we now take for granted. In particular, Jeremy Oppenheim, Roman Krznaric, Kate Raworth, Tristan Hanson, Nigel Kershaw, Kevin Riches, Bernadette Wightman, James Cameron, Adam Tooze, Joseph Stiglitz, David McWilliams, Martin Wolf, Dennis Snower, Simon Wren-Lewis, Tabitha Morton, Austin Rathe, Bess Mayhew, Maurice Biriotti, Aidan Regan, Stephen Kinsella, Karl Whelan, Ed Brophy, Carolina Alves, Sony Kapoor, Anne Richards, Michael Collins, Mason

Woodworth, Shamik Dhar, Nahed Sarig, Gautam Samarth, Alex Seddon, David Knee, Graham Mason, John William Olsen, Alex Araujo, Dave Fishwick, Tony Finding, Alex Houlding, Jenny Rogers, Craig Simpson, Maria Municchi, Marco Spizzone, Clare Patey, Adam Ross, Alev Scott, Jen Danvers, Tim Goldfarb, Daniella Waddoup, David Snower, Antonia Burrows, Amy Porteous, Rose La Prairie, Stewart Douglas, William Newton, Victoria Buhler, Eoin Condren, Patrizio Salvadori, Oliver Sawers, Sam Sawers, Theresa Thurston, Louise Newman, Lidia Lonergan, Richard Jolly, Chanya Button, Jonny Howard, Grace Ofori-Attah, Jessie Newman, Arthur Worsely, Erin Young, Alex Brunicki, Harry Simpson, Daniel Mytnik, James Hanham, Henry Bartlett, William Archer, Ida Bafende, and Alex Peirce. We have also had many fruitful discussions with politicians from many countries and political parties.

We had the opportunity to road test some of our ideas at a conference on climate change policy organised by the Centre for European Reform. Thanks in particular to the organizers, Charles Grant, John Springford, and Christian Odendahl, and all the participants, in particular, Anatole Kaletsky, Jean Pisani-Ferry, Stephen King, Martin Sandbu, Daniela Gabor, Catherine Fieschi, Angel Uribe, Jeromin Zettelmeyer, and Shahin Vallee. Megan Greene has also been an invaluable interlocutor at many points.

The team at Positive Money Europe, and Stan Jourdan in particular, have been great collaborators in recent years. Stan is an independent and creative thinker, and has influenced many aspects of our thinking. Thanks also to Maria Demertzis and Gregory Claeys at Bruegel, Thomas Fricke and team at the Forum for a New Economy, Jan Bruin from Brainwash, and Aimee Ling, Lizzie Cho, Stewart McCure from Nova.

ACKNOWLEDGEMENTS

We particularly want to thank Angus Armstrong, for his contribution to our thoughts around global policy solutions, and Sebastian Leape, Kathryn McLeland and Udayan Guha, for furthering our thinking on green mortgages. Thanks also to Michael Burleigh, Frank Van Lerven, Carys Roberts, Torsten Bell, Simon Tilford, Jonathan Hopkins, David Andolfato, Tony Yates, Philip Coggan, John White, Graham Pattle, Frances Coppola, and Joe Wiesenthal – for thought-provoking exchanges, sometimes on Twitter.

Agenda is an innovative and flexible publisher. A huge thank you to Steven Gerrard and Vicky Capstick for all their hard work and tolerance of demanding authors.

It goes without saying, that almost no one who has provided feedback, inspiration or help, agrees with everything we have written – many disagree strongly!

Maia, Gina, Rafaele, Corinna Salvadori Lonergan and Gladys Tomarong have provided consistent support, inspiration, and periodic distraction.

Corinne Sawers and Eric Lonergan

Introduction:
a failure of the mind

"If our poverty were due to famine or earthquake or war –
if we lacked material things and the resources to produce
them, we could not expect to find the Means to Prosperity
except in hard work, abstinence, and invention. In fact, our
predicament is notoriously of another kind. It comes from
some failure in the immaterial devices of the mind, in the
working of the motives which should lead to the decisions
and acts of will necessary to put in movement the resources
and technical means we already have. It is as though two
motor-drivers, meeting in the middle of a highway, were
unable to pass one another because neither knows the rule
of the road. Their own muscles are no use; a motor engineer
cannot help them; a better road would not serve. Nothing is
required and nothing will avail, except a little, a very little,
clear thinking." John Maynard Keynes (1933)

There are many great books on climate change, but very few
address what we need to do, in a convincing way. Personal "to-
do" lists suggesting we eat less meat, buy an electric vehicle,
and fly less, seem trivial relative to the challenge of climate
change. We need serious policies at the individual, corporate,

national and global levels, which can rapidly accelerate the pace of change.[2]

There is also a lack of clear thinking. A collapse in global CO_2 emissions does not require a complete change in how we live our lives, or a huge increase in taxes. It requires ending our dependence on fossil fuels. This presents us with an enormous investment challenge. Most of the world's electricity, transport, buildings and industry are powered using fossil fuels. We need to power them with renewable energy. If electricity is generated with zero emissions, and transport, buildings and industry are all powered by electricity, global emissions would fall by around 75 per cent. This is the greatest reallocation of capital in human history.[3] We should be doubly angry if we don't rise to the challenge – investment is the means by which we become wealthier.

The quote above from the economist John Maynard Keynes was written during the Great Depression. Keynes observed that mass unemployment was a "failure of the mind". History has proven him right. When the world was hit by the global pandemic, almost all countries followed the sort of policies of government intervention that Keynes advocated in the 1930s, and economic collapse was averted.

The same is true for our response to climate change. There is no good reason for our failure to act faster and far more effectively. We have the technology, by and large, to replace fossil fuels, and the financial resources to pay for the investment. It will require a huge amount of work, but the economic opportunity is immense. Almost all of the global population benefits from a transition to clean renewable energy and widespread electrification. This is analogous to building our transport system, our telecommunications infrastructure, or the Internet.

History has taught us that investment on this scale improves our quality of life and creates wealth.

So why isn't it happening? Despite all of the research on climate change mitigation, policy-makers have not yet cracked a set of policies to deliver decarbonization on the timescales we need.[4] The absence of effective policies is the major gap in our thinking. There are books that scare the hell out of us. Books such as *The Uninhabitable Earth* by David Wallace-Wells, which maps out the risks of inaction, and has helped to develop a consensus on the scale and urgency of the problem.[5] There are others that explain where our efforts should be focused. Books such as Bill Gates's *How to Avoid a Climate Crisis*, clarify where most emissions are being generated, and assess new technologies and scenarios for emissions reduction. There is a wealth of important research on climate science and potential actions for mitigation published by international organizations.[6] Alongside this, powerful work and deep thinking has also been done on the structure of capitalism, the unspoken rules, incentives and beliefs, which shape that system: books like Rebecca Henderson's *Reimagining Capitalism*, and *Doughnut Economics* by Kate Raworth.[7] These are all invaluable contributions that have helped to build knowledge, consensus and action over time. But where is the manifesto which shows us how to do it?

Bill Gates ends his otherwise insightful book by recommending, among other things, that we stop eating burgers. This is the right thing to do, yet entirely unsatisfactory. Human beings don't change their behaviour in response to a list of action points at the back of a book. We need a new set of radical economic policies at the national and international level that will deliver accelerated decarbonization and we also need a realistic view of how individuals and society changes, how institutions such as businesses change, and how the government can tilt the playing field firmly in our favour.

This is the goal of *Supercharge Me*. We propose a set of measures that can accelerate this process of global economic and social change, and in a way that ensures most people will experience immediate benefits. These policies stand a very good chance of being implemented, and working. They should also give us a sense of optimism, a sense that for most people the gains are greater than the sacrifices. This is important. Part of a viable political strategy is winning hearts and minds.

Supercharge Me is grounded in relentless realism about how governments, businesses and individuals actually behave. It's also based on what has worked so far: positive incentives and hard regulations. Our ability to manoeuvre in this crisis is limited by human and corporate psychology. People only change behaviour if the alternative is cheaper, better, or their friends are doing it. Businesses only change behaviour if they can make more money, or if it's a legal requirement. Policies only work, and stick, if they are effective, simple and non-partisan. An astonishing amount of climate mitigation energy is going into tactics that ignore these fundamental principles.

Our goal is also to lay out a plan for tackling the climate crisis that is accessible to any reader. This means we have prioritized clarity and simplicity of expression. Some readers who have expertise in the topic will want more, so we have provided technical detail and extensive references in the endnotes. This book is also a dialogue. We think this makes it more engaging, but it also reflects our different areas of expertise: in sustainability and systems change (Corinne) and in economics and policy (Eric). The two voices reflect our different perspectives and allow us to invite the reader to be part of our conversation.

The first chapter defines "supercharging" and introduces the concept cf "extreme positive incentives for change" (EPICs). This w_ll be central to our argument. EPICs are a practical theory of policy-driven rapid change. In the context of decarbonizing our economy, the evidence is compelling that extreme positive incentives are highly effective. They have already delivered a transformation in the pricing and adoption of solar power, and rapid take-up of electric vehicles in parts of China, Scandinavia and the United States.[8] EPICs combine two key ideas: positive incentives are more effective than negative incentives and extreme incentives can trigger rapid changes in behaviour. If a plant-based burger is only priced 3 per cent lower than a regular burger no one will change behaviour, but if it's 30 per cent cheaper who wouldn't give it a try? By contrast, extreme negative incentives, such as draconian carbon taxes, can trigger substantial resistance and as a result fail to happen. We shall evidence this in detail in Chapter 1, and then exploit the conclusions throughout the book. Chapters 2 and 3 detail the profound changes occurring in the corporate sector and in financial markets An extraordinary cultural shift has occurred in the last few years. Why do we care? Because it frames the policy opportunity. If policies are designed smartly, financial markets and corporate culture are primed to deliver a disproportionate response. The barrel is loaded.

We are both professionally close to these rapid changes. We are also both concerned that despite the unambiguously positive opportunity emerging, the train may come off the rails, and slow progress. There will be greenwashing scandals – sustainability after all is a marketing opportunity like none other – and the bad eggs will discredit broader progress. Both risk compromising the supportive public sentiment that is critical to a successful transition.

Throughout this book we repeatedly return to first principles. In Chapter 4, we discuss the power of individuals. Don't expect a to-do list with tips on ride-sharing. We draw on the latest thinking in social and behavioural psychology to inform our understanding of how individuals actually change their behaviour. We need to shift social norms. Otherwise we are wasting our time. Motivated minorities have always been agents for social revolutions, and activist-individuals have never been more powerful than in the age of social media.

Chapters 5 and 6 focus on policies at the national and international level. *Angrynomics*, which Eric co-authored with Mark Blyth, proposes the criteria for "smart" policies.[9] First of all, they have to be effective. This sounds obvious, but it needs to be said. Economists often design policies based on what is "optimal" or "efficient". But the policy which delivers the best outcome in the textbook may fail in reality. Many policies also fall into the class of making us feel better, but having marginal impact. With climate change it is even more important that policies have a high probability of maximum impact. Ideally, proposals should also be non-partisan. This has the advantage of being politically viable, and protects critical changes from being reversed with the electoral cycle. Policies must still be radical, because we require an overhaul of our energy system, so the policies we propose in Chapters 4–6 are extreme. Finally, a powerful policy should be as simple and clear as possible. EPICs are good examples of this, but we shall also apply this standard to regulations and taxes. We are only interested in regulations that are relatively simple and that have a very high probability of success, what we call "smart regulations".[10] It is a helpful coincidence that simple and effective policies also reframe the debate. That is an area where we really hope to succeed. A key purpose of this book is to change perceptions of the economic consequences of transitioning to a green economy. Hopefully, in the space of a few hours, readers will have a very clear sense of what the problem is, that it is tractable, and that it can be solved. They will also be armed with three or four powerful and simple policy prescriptions to fight for.

Chapter 5 deals with national policies. We show how EPICs can transform our energy infrastructure, and corporate behaviour, not in 20 or 30 years, but in the next five to ten years. Chapter 6 outlines both bilateral and multilateral policies, and considers, for example, how developed countries can help poorer nations, such as India, accelerate their momentum towards sustainable electricity, or how the global steel industry, which emits greenhouse gases almost eight times that of Britain, can transition to produce green steel.[11] Chapter 7 ties up some loose ends, diving deep into the critical topics of methane emissions, deforestation and carbon capture.

We wrote this book because we are witnessing not just a climate emergency but a failure of the mind. Preventing global warming from breaching 1.5°C requires a rapid wholesale shift in how our economies are working.[12] What is absent from the public debate is clear awareness that we have the technology and the financial resources to transform our economies in a way where most people will experience tangible improvements in their quality of life.

1

Supercharging: what is it?

"A striking difference between commitments under NATO, and commitments under the Kyoto Protocol is the difference between commitment to actions and commitment to results." Thomas Schelling

ERIC: What is "supercharging"?

CORINNE: Supercharging is an action plan to accelerate the decarbonization of the global economy. It is based on realistic assumptions about the motivations of individuals, businesses and governments. Serious targets are now embraced by Europe, the United States, China, and India. The goal is to prevent the earth's temperature from rising more than 1.5°C.[1] A huge amount is happening, but the current trajectory of decarbonization falls far short of this ambition.[2] The striking omission is a coherent framework for action.[3] Supercharging is not about targets, but actions. At every point, we ask: *What can be done to reduce emissions rapidly, based on a realistic view of how individuals, businesses and politics work?* Highly complex global carbon taxes, or trivial advice on lifestyle changes, don't cut it. Taxes which work in the textbook often fail in reality, global lifestyle shifts don't even appear in theory. We need to refocus every area of policy to replicate the extreme successes in pockets of the world where rapid growth in sustainable energy is occurring, and where consumers are voluntarily switching purchasing decisions en masse. There are already successes in parts of China, India, Europe and the United States which prove that we have the technology and the resources to collapse emissions.[4] If these models are replicated, if we supercharge the global economy, we will not only have sustainable economies and cleaner air, there are also very good reasons to believe that we will have higher incomes.

ERIC: History shows that technology-led transformation of our economies greatly improves the quality of our lives and our living standards.[5] The green transition should be no different. This is not idealistic. There will be some losers, particularly owners of fossil fuel assets. Fortunately, in most of the world these assets are largely in the hands of three groups: thugocracies, theocracies, and a fraction of the "1 per cent".[6] The overwhelming majority of the planet need not worry about the value of these assets falling towards zero.[7]

CORINNE: In order to supercharge the world, we need a clear diagnosis of the problem, a theory of change, and very smart policies. How would you summarize the problem?

ERIC: Climate change is sometimes described as a "wicked problem". In an interview on YouTube, the behavioural psychologist, Dan Ariely, argues that if you wanted to design a problem that humans would not care about, it would be global warming.[8] It makes for depressing listening. Firstly, humans are hardwired to prioritize the immediate over the future. When we don't see things happening, we find it very hard to be motivated. Secondly, we often pay more attention to the plight of one person than that of large numbers of people. Ariely cites the fact that CNN devoted more coverage to the story of baby Jessica stuck down a well, than the humanitarian crisis in Rwanda. To make matters worse, prevention is much less motivating than a disaster that has already happened. When we have to make short-term sacrifices for long-term benefits, we convince ourselves that we don't have to make these changes.[9] This might explain why until recently crank climate deniers, were often concentrated in geographies where the economic consequences of decarbonization were more severe. It also explains in part why we hear lots of "good" reasons for inaction: "I can't do anything about it, it's all due to Chinese coal consumption", or "What's the point in me not taking flights abroad, if everyone else is?".

CORINNE: Ariely also argues that if we want to tackle climate change we need to exploit these motivations, rather than ignore them, and fail. That is precisely how we should stress-test policy frameworks and recommendations. This is why supercharging is not just essential, but also likely to be much more effective. People are more motivated if they can witness quick results, and if the problem is perceived to be tractable. They are even more motivated if they can see quite rapid benefits to them. This has to be the context in which we tackle climate change. It is the litmus test of a successful programme.

ERIC: So the first challenge is to reduce an extremely complicated process – decarbonization of the global economy – to a set of relatively clear objectives. As someone who has studied this problem for over a decade, can you help?

CORINNE: I am going to do this in two stages. First, I am going to give a clear overview of where we stand. Then it's going to get really interesting. By following the strategy we advocate, the problem actually gets more tractable through time. If three or four things fall into place, all of which are highly amenable to supercharging, around 75 per cent of emissions can be eliminated on timelines far shorter than the most ambitious existing national goals.

ERIC: Talk me through the first stage, where we currently stand.

CORINNE: There is a clear breakdown of the main sources of greenhouse gas emissions which are causing global warming. Bill Gates frames it neatly around five activities.[10] To paraphrase: switching things on, making things, getting around, growing things, and heating or cooling buildings. We can roughly quantify the contribution of each of these. Switching things on – electricity – accounts for 27 per cent of global emissions. Making things, such as manufacturing steel, cars, clothes and computers, adds another 31 per cent. Transport contributes a further 16 per cent to global emissions. Growing things – agriculture – including crops and livestock, makes up 19 per cent. The remaining 7 per cent or so of global emissions are generated by heating and cooling buildings. These numbers are approximations, but give an accurate picture of our priorities.[11]

ERIC: That's clear. So why do matters get simpler through time?

CORINNE: In a snapshot, the problem collapses to one dominant factor: electricity. If electricity generation is close to 100 per cent sustainable, and we run as much as possible off the grid, emissions collapse. Electrifying everything won't get us to a completely decarbonized economy, but it can achieve the lion's share. Gates's five areas which contribute to global emissions don't have to be targeted independently. We now have the technology to use electricity to run almost all road and rail transport and regulate the temperature of buildings.[12] With current technology we can provide close to 50 per cent of the industrial sectors' energy needs with electricity, and this share is rising rapidly with innovation.[13] This leaves "growing things" (agriculture), and some intractable areas of transport, such as air travel, as the significant areas that will require different strategies.[14] A targeted policy that converts electricity generation to sustainable sources, and powers all ground transport, heating and the majority of manufacturing with electricity, has the potential to reduce emissions by 75 per cent.[15] Electrification of these sectors, and transitioning to renewables are both highly suited to supercharging. There are very important caveats and complexities, which we shall come back to. But first we need to deliver the punchline. It is immediately clear that in a fundamental sense Dan Ariely is wrong.

ERIC: Explain.

CORINNE: You're the economist...

ERIC: Ariely accurately describes human motivations, something which we take on board. But he misconstrues the problem of climate change. He presents the problem as if we are trying

13

to wean kids off chocolate. He assumes, like many do, that we all have to make unpleasant changes to our behaviour now, in order to benefit other people in the distant future. The diagnosis you present is completely at odds with this perspective. The problem you have described is primarily one of investment, not restraint or abstinence. It requires building new power plants using solar and wind to transform our electricity generating capacity. Now this is concentrated with a relatively small number of actors, whose investment decisions are highly sensitive to financial incentives.

CORINNE: That's right. Now we need to explain in more detail the process of supercharging. Let's start by clarifying our thinking around timelines. How can countries get close to targets of 75 per cent or 100 per cent sustainable electricity generation within five to ten years?

ERIC: There are significant practical challenges, but first we need to understand why the economics of sustainable power generation lends itself to policy-induced acceleration. When we think about building new power plants, laying transmission cables, and building an electricity grid, we tend to assume long time lags to completion, decades rather than years. Another aspect of our good fortune is that the two critical clean technologies, wind and solar, can be installed much more rapidly than legacy power generation, such as coal, or controversial sources of energy, such as nuclear. Smartly combined, the private and public sectors can build solar, wind and other renewables at scale and speed.

CORINNE: To give some context, a 50-megawatt (MW) wind farm can be installed within a year.[16] With 50 MWs of electricity you can power thousands of homes.[17] Even in developed economies with relatively high population densities, widespread Nimbyism and heavy planning regulations, huge projects can be delivered from start to finish within five years.[18]

ERIC: Do we have any examples of this being done at scale on these timelines?

CORINNE: There are many examples, some from China and India, which may surprise people and which we shall describe later. But if we restrict ourselves to the developed world where planning constraints are greatest, UK offshore wind is a very good example of what we are advocating.[19] The UK originally planned to double its renewable energy capacity by 2030. Under current plans for offshore wind, renewable capacity will double by 2026. This will take installed wind and solar capacity to around 64 gigawatts (GW), compared to a *total* installed capacity in the UK of 76 GW in 2020.[20] That is supercharging, and has happened by applying the policies and principles we describe.[21]

ERIC: How much of global electricity generation could be provided by wind turbines?

CORINNE: Assessments of feasibility based on our technological capacity and geography by climate scientists at Harvard, using reasonable assumptions, suggest that wind power alone could power five times global energy consumption.[22] To be clear, that is not five times global electricity usage, but *total energy usage.* If we include solar, which is now the cheapest form of electricity generation, many of the limitations of wind power are mitigated, and our capacity is further increased.[23]

ERIC: If we look at the globe, the two biggest greenhouse gas emitters are the United States and China. How suited are these continental-scale nations to alternative sources of electricity generation?

CORINNE: Here, again, we are actually very lucky. Both China and the US are in the top five geographies where wind, solar, or a combination of the two, can be deployed at scale. Ten times the electricity needs of the United States, for example, can be met from wind alone.

ERIC: So the biggest geographies which need to be transformed, the US, China and . . . what about India?

CORINNE: India too, in fact. India has a proven capacity to create alternative sources of power generation at scale. India already has the fifth largest installed renewable capacity in the world.[24] It is the only country in the world to host a solar-only airport, in Kerala. Of course, it still has a power system heavily reliant on coal. The challenge, when we get to global policies, is how to incentivize all nations such as India to move much faster, and tackle the huge resistance from their incumbent polluting sectors. It is worth repeating that the losers in a well-managed green transition should not be the general population, or consumers, but the owners of existing polluting assets, and in some geographies, the political structures that these support.

ERIC: Ok, so the most significant geographies happen to be near-optimal from the perspective of solar and wind power. The lead times for capital expenditure in solar and wind are far shorter than for traditional power generation.[25] There is no physical barrier to building wind and solar power in two to five years. So why hasn't it happened? Where are the blockages?

CORINNE: To a significant extent it *is* happening. In 2020, 82 per cent of new electricity capacity globally was wind or solar.[26] There are practical challenges around electricity transmission, and the obvious fact that wind and solar vary by season and time of day, which requires accelerated investment in storage, and increased capacity. But these are not the real obstacles. The main reason things aren't happening fast enough is because the economics don't work, the returns to investing faster and at a greater scale are not yet there. Planning is too slow, and vested interests block progress.[27]

ERIC: So supercharging involves changing the economics, calling out vested interests, and identifying the formula at the heart of the successes?

CORINNE: Exactly. For example, why has the UK not replicated the model it uses with huge success for offshore wind to the solar industry, or electric vehicle take-up? Why is India not harnessing the same principles which are creating extraordinary growth in solar capacity, to repurpose its coal capacity, or convert its steel industry to green steel? Why are some US states targeting close to 100 per cent sustainable electricity in ten-year timelines, and others are taking 30 years? And why are none doing so in five years?[28]

ERIC: So the challenge is clear, how do we incentivize and motivate the global community to accelerate the transition?

CORINNE: I think it helps to break down what needs to be done into three categories of problem. There are problems where we have established technology, and the obstacle is primarily economic. There are problems where we need new technologies, and there are those which require individuals to change their behaviour, the types of problem Ariely focuses on. This framing really helps, because each type of problem requires a different approach. It will also help when we look at supercharging the individual, the nation, and the world. There are some big wins, which we need to identify, and accelerate. We can then focus resources and buy time for the difficult jobs. I will explain this approach in more detail shortly, with clear examples. But before doing so, we need to explain extreme positive incentives for change (EPICs), which may be one of our most important ideas. EPICs are behind the most successful developments in tackling emissions of the past 15 years, such as the collapsing cost of solar energy, the boom in global investment in wind power, and the uptake of electric vehicles. The effectiveness of EPICs needs to be recognized and they should be adopted more widely. Investment in electricity generation, in transport, in buildings and in manufacturing, is particularly sensitive to an EPIC. So, Eric, what is an extreme positive incentive for change?

ERIC: In this context an *incentive* is a financial benefit encouraging a change in behaviour. A *positive* incentive is where the effect makes you better off relative to the status quo, not worse off.[29] *Extreme* refers to the fact that moderate incentives often have no effect. We need incentives to make a material difference, so we really notice it, and it affects our behaviour.

CORINNE: Why do you think EPICs are the central mechanism for supercharging? There is, for example, a huge focus in climate policy-making on carbon taxes. Do taxes have a more potent role within the context of EPICs?

ERIC: The idea behind EPICs draws on a combination of microeconomics, behavioural psychology and political science. It exploits the economic features of research and development, and economies of scale. The response of businesses and individuals to large positive incentives is often predictable, and can be very pronounced. This is often not the case with other types of policy, such as taxation. The huge changes in corporate and consumer behaviour that we've seen in recent decades, such as the emergence of companies like Tesla, the explosion in solar power generation in developing countries like India, the mass adoption of electric vehicles in parts of Scandinavia, have involved positive financial incentives for innovation and the adoption of new technologies.[30] Positive incentives create support instead of opposition. The emotional and behavioural response of consumers is completely reversed if they can get a cheaper mortgage because they insulate their home, as opposed to facing new taxes if they don't. Instead of feeling penalized, or financially worse off, something people will naturally resist, positive incentives get people on-side and change outcomes.

CORINNE: Let me respond with a challenge. In the world of climate policy, the framing focuses on the fact that those responsible for CO_2 emissions are not paying for them.

Economists call this an externality, and the way to deal with externalities is to tax them. This also appeals to our sense of justice. Extreme positive incentives for change sounds counterintuitive against that backdrop.

ERIC: It is worth spending some time on the concept of an "externality". The idea is not new. The first mention is found in the writing of the great Scottish philosopher, David Hume.[31] Hume discusses the idea in the context of social institutions such as language, law and money. He observed that the effects of human activity can be external to its intentions or focus. If we individually obey the law, or engage in exchange, there are huge collective benefits which occur to society which are not inherent in our individual intentions. The relevant application in environmental economics is the idea of pollution, a cost associated with trading goods and services which is not captured in the price.

CORINNE: The standard approach has been to try and compensate with taxes, to try and include the external cost of pollution in the price of carbon.

ERIC: Externalities are relevant to "fairness", the idea that the polluter should pay for the damage they are doing. Taxes directly targeted at emissions can succeed in raising large revenues for the state, but it's not clear that this delivers the rapid reduction in emissions that we need. There are simple reasons for this. For example, we rarely significantly adjust how many miles we drive in response to changes in petrol prices.[32] By contrast, EPICs focus on substitutes. When thinking about rapid behavioural change, focusing on substitutes is more important than accurately taxing externalities.[33] If substitutes don't exist, very little change occurs. So taxes on carbon need to be combined with EPICs to affect behaviour at scale and speed.

CORINNE: There are also important non-economic reasons why EPICs are more effective. Humans respond very differently to immediate gains, than to costly requirements to change. If I say, "We will only stop global warming if you don't travel abroad, buy fewer clothes, and stop eating meat" most people will feel deflated. Much worse, if we ask developing countries to limit their potential for economic growth and constrain their people's standards of living, we will understandably meet with resistance. Alternately, if I say that an electric vehicle will be 15 per cent cheaper to buy, air will be cleaner, you will have a lower electricity bill and lower mortgage interest rates as a reward for going electric, and the transition will create more jobs and rising wages, the behavioural response is entirely different.

ERIC: We have to take these facts on board. The dominant political strategies for achieving decarbonization can have the opposite consequence to what we are intending. It can dissuade people from the cause. And they aren't even reflecting the economic reality of the transition.

CORINNE: I can see the case for positive incentives over negative incentives, particularly in light of the psychological challenges that Ariely describes. It has been shown that people are more willing to believe evidence or support a cause if it is in their financial interest or not too inconvenient for them.[34]

21

ERIC: Agreed.

CORINNE: Talk me through why EPICs need to be "extreme"?

ERIC: Our response to incentives is not linear. The more extreme the incentive the more powerful the behavioural response. There is no point in the government setting up schemes to provide small savings from installing an electric heating system in residential property. Most people will not incur a substantial inconvenience for a small, slow, cumulative benefit. For exactly this reason, widespread attempts to tackle heating efficiency in residential properties have failed. When we come to discuss supercharging the nation, we shall show how to use EPICs to change this. [Spoiler alert: if your neighbour is saving 30 per cent on their energy bills, getting a lower interest rate on their mortgage, which rolled up the cost of installing an electric immersion heater or heat pump, the probability that you will copy them is high. It's that simple.]

CORINNE: The greater the incentive, the higher the likelihood of change. That makes sense. I'm keen to discuss whether this holds equally for consumers and corporations. But before we address this, can you explain how taxes and regulation fit into this picture of extreme incentives that we advocate.

ERIC: Overwhelmingly, the evidence points to sequencing two successful strategies for behavioural change – EPICs and smart regulations. We need to be clear that there is a role for both, and carbon taxes need to work within this context. Our strategy is to supercharge and front-load with EPICs, and then tax and regulate where possible and where effective.[35] The sequencing needs to work this way for two reasons.[36] First we need to promote substitutes for polluting goods. There is no point in taxing a petrol car if a competitively-priced electric option is not available. This principle needs to be applied in every sector. Focusing on making the green substitute much cheaper requires a substitute to be found and scaled up. The second factor is political. We need significant competing vested interests in the green sectors – we need the wind farm and solar industries to lobby against oil. The same could be said of the wider population. We need a large constituency to support regulations in areas where they are not hugely affected. We cannot simply rely on a mass conversion to green ethics.

CORINNE: If we look at the success stories, EPICs have been central to scaling EVs, solar power, battery technology, and wind energy. Huge subsidies triggered rapid innovation, and ultimately big breakthroughs on cost. This is the key to dramatic, rapid change.

ERIC: We should briefly describe some huge successes delivered by EPICs in combating the climate crisis.

CORINNE: Solar power is a good example. The original breakthrough in solar technology occurred in 1940, when a researcher at Bell Labs shone a bright light onto a rod of silicon creating a current between the electrodes at the rod's ends. By 1954, Bell Labs had developed the first solar battery.

ERIC: What have we been doing for the past 70 years?

CORINNE: EPICs are the key to understanding why it is only

in the past two decades that solar panel prices have collapsed and global demand taken off. New technologies are rarely economically viable at inception. Markets need to be created, and when demand accelerates, unit costs of production fall. In the case of solar panels, Japan, Germany and China all deployed EPICs to create today's global market.[37] Between 1970 and 2000, the global price of solar power fell by around 90 per cent, due to Japanese EPICs, under its so-called "sunshine programme", which started after the second oil crisis in 1979.[38] Due to large and consistent incentives to manufacturers and households over a decade, Japan became the world's leader in solar manufacturing. Mistakenly, Japan abandoned this policy in the mid-2000s, and the baton was taken up elsewhere. In Germany the solar sector was kick-started by a coalition government including the Green Party. By offering fixed-price contracts for 20 years to solar power generators, the industry boomed. An EPIC changed the world. Between 2004 and 2010 the global market for solar panels grew 30-fold.[39]

ERIC: This is an example where an entire industry has been transformed through EPICs. Are there examples of consumer behaviour changing rapidly?

CORINNE: EPICs often need to be applied both to supply and demand, particularly in the early stages of creating a market. We need to bear this in mind when we think about creating markets for green steel and cement. But there are many areas where the supply is available and EPICs are being applied directly to change consumer demand. Norway's extraordinary success with electric vehicles is an illustration of how EPICs can rapidly change individual behaviour. By 2020, the list price of an electric VW Golf in Norway was lower than the petrol model, largely due to tax exemptions. In addition, Norway had introduced the 50 per cent rule, an EPIC. Under this rule, owners of electric vehicles pay no more than 50 per cent of the price charged for fossil fuel cars on ferries, public parking and toll roads. EVs are also exempt from road tax, and have access to bus lanes. The annual savings amount to several thousand dollars. Is it any surprise EV sales in Norway exceeded 90 per cent of total car sales?[40]

ERIC: It's actually surprising that anyone bought a petrol vehicle. The logic is compelling. Make the incentive positive, make it extreme, and people respond. Contrast this with a carbon tax. The UK has one of the highest levels of fuel duty in the developed world, and yet the penetration rate of electric vehicles is unremarkable and behind that of Germany and France.[41] There is even a consumer organization fighting fuel taxes.[42]

CORINNE: There are two points to remember from this contrast, Norway's policies have succeeded in dramatically changing behaviour, and public perceptions of the consequences of climate policies are very different to those in the UK. This is the difference between using a simple carbon tax and an EPIC. We need to be very clear. Taxing fuel and fossil-fuel vehicles is necessary, but it doesn't deliver public support or behavioural change on the scale and at the speed that is required. A carbon tax is most useful when it is a complement to an EPIC.[43]

ERIC: Perhaps unsurprisingly, China's most successful electric car capital, the southern city of Liuzhou has done something similar to Norway, if not quite as extreme. In 2020, sales of EVs in Liuzhou accounted for 30 per cent of the total.[44]

CORINNE: I have one concern. This is a compelling approach to rapid behavioural change, but if all incentives are positive, the government is losing tax revenue. How do we pay for EPICs?

ERIC: We shall discuss this in more detail later, when we also consider how EPICs can be applied to the central challenge of converting global electricity to wind and solar. The key point is that we have no choice, we have to use EPICs. They can be implemented through monetary and fiscal policy. In the case of infrastructure, EPICs are substantially self-funding, as we shall explain. If we do need to raise taxes to cover lost revenue, or we need to raise taxes for other reasons, we should approach this independently. There are many areas of our economy which could be taxed more efficiently.[45]

CORINNE: Ok, we have outlined the strategy. Make electricity generation green. We have the technology, and capacity can be built at scale and speed. The areas of the world which create most emissions are also ideally suited to wind and solar power. If we rapidly green global electricity, we must electrify transport, manufacturing, buildings, and as far as we can, agriculture. This is our high-level challenge. Now, I want to return to how we broke down the problem earlier into three challenges: where the obstacle is primarily economic, where we need new technologies, and where we require individuals to change their behaviour. Let's call these "simple maths", "mini-Musks" and "herding sheep". First, "simple maths" – these are problems where a technological solution is already being widely used, is scalable, and accelerating the process is largely a question of pricing the incentives correctly. It's "simple maths" because if a certain numerical threshold is met, a huge change happens quite rapidly. In these categories, EPICs are the route to supercharging.

ERIC: This goes to the heart of electrification – if we get the EPICs right, the private sector will deliver accelerated investment spending in wind and solar. Electricity infrastructure can be rolled out globally and fuelled by renewables.

CORINNE: Transport, other than air travel, is also a "simple maths" problem. Follow the Norwegian model and passenger road transport could be fully electrified within a decade.[46] Not only that, drivers and citizens will be happier for it. Of course, Norway illustrates that behavioural change in transport is a numbers game: make electric vehicles 10 or 20 per cent cheaper than the petrol equivalent. But even Norway hasn't gone far enough. We also need to replace the stock of fuel-based cars. An EPIC-based approach needs to be part of a national and global plan to replace internal combustion engines, particularly for those on low incomes. Despite all the progress, only 1 per cent of the global stock of cars are electric.[47]

ERIC: What about home-heating and making things, like steel and cement?

CORINNE: Despite its complexity, many aspects of electrifying and insulating buildings are "simple maths" problems – air conditioning, for example, or installing immersion heaters. Steel may now have become a "simple maths" problem, too.

Technologies exist for green steel, but they are not competitively priced. Incentives are needed to reverse this, which we shall discuss. By contrast, "mini-Musks" are problems where we need technological breakthroughs, and a willingness to throw billions of dollars at research and development, with highly uncertain returns.

ERIC: What are the main areas where we need more Musks?

CORINNE: Cement and aviation would be top of my list, as well as agriculture. Food is partly a "mini-Musk" problem, partly "herding sheep".

ERIC: Explain the latter.

CORINNE: There are genuine consumer behaviour challenges where EPICs may not suffice – activism targeted at social stigma may be needed, to cancel consumerism. I call this "herding sheep". These ideas come to the fore when we discuss supercharging the individual. When it comes to changing human behaviour, there are areas where people are genuinely indifferent, for example, consumers don't really care what type of engine is in their car. When it comes to changing food habits, it's much more complex. Food is firmly embedded in our cultures and customs. We have deep, emotional associations, often linked to upbringing, family and ritual. That's why we struggle to give up mince pies and cream at Christmas or mithai at Diwali, despite the pursuit of abs.

ERIC: Can innovation and EPICs help with the food system challenge?

CORINNE: Absolutely, and so can activists. But innovation is vital. Plant-based meat is a significant breakthrough, and requires far less sacrifice from the consumer. They can still eat a burger, and the experience is basically the same.

ERIC: I disagree with that, but I'm confident the technology will continue to improve. Before we wrap up with a consideration of winners and losers, it might be helpful to outline how we intend to structure our discussion and where it will take us. We plan to apply our framework to global policy, national policy, corporate behaviour, greening finance and individual action.

CORINNE: That's right. We shall show how each can be supercharged with EPICs. And where other strategies are needed, we have to be much more focused in deploying them. Importantly, EPICs can create a huge positive tailwind. It's important to win people over by delivering upfront benefits.

ERIC: Great. So let's quickly discuss winners and losers.

CORINNE: So far, this all seems too good to be true. My inner puritan feels that there have to be major costs. How can all these subsidies be good for the nation's finances?

ERIC: This is an area where economic terminology has not helped in framing the debate. Most of what we need to do to get to net zero involves capital expenditure, investing in infrastructure. Now, the return on most of these investments will exceed our funding costs – so these investments will result in creating assets. That means we are better off, independently of

29

saving the planet.[48] I don't think this is a "cost", nor do I think that is what most people understand as a cost.

CORINNE: If I take out a mortgage and buy a property, which I rent out for more money than the interest rate on my loan, I am wealthier. Is that the correct analogy?

ERIC: Yes, exactly. People read a report on climate change that says it will cost hundreds of billions or trillions of dollars, and get depressed. But what they don't realize is that it won't actually cost us anything: we will borrow and invest, creating valuable assets and jobs. Economists, however, think in terms of "opportunity cost". In other words, we could have invested in something else. Now this is true as a matter of logic, but very misleading in reality. At the moment, there is chronic under-investment in the global economy. By increasing investment we can create huge value for society. This is a fact of arithmetic, a result of historically low interest rates across the world. In theory, we could build more schools and hospitals instead of building more electricity infrastructure. In reality, however, we are not. So as a practical matter, this is not a sacrifice.

CORINNE: Does this change if interest rates rise?

ERIC: It could do, but they would need to rise a lot. The key point, however, is that these huge estimates of the "cost" are very misleading. In purely financial terms the investments we are trying to accelerate, which do amount to trillions of dollars, make us collectively wealthier than we are today, with the exception of a small minority of losers. It is critically important for people to understand this.[49]

CORINNE: Nicholas Stern, author of the 2006 landmark report on the economics of climate change, makes a similar point in his more recent book, *Why Are We Waiting?*, that there is a positive economic case for investing in climate mitigation.[50] What you are emphasizing here, is that aside from mitigating the economic costs of climate disaster, an industrial revolution – which is what decarbonization is – will result in higher living standards.[51] But there must be some losers. Talk me through who they will be.

ERIC: The debate on tackling climate change repeatedly presents as a negative, but the core of the solution is unambiguously positive for almost all members of society. And I don't just mean it's a benefit because we prevent the climate crisis. The *wealth* of our society will rise, and the losses will be concentrated with a very small minority. Investment creates wealth if the financial return exceeds the cost of borrowing, which it is almost certain to do when long-term interest rates are negative in real terms. And who owns the assets whose value we are destroying? Fossil fuel assets, owned by oil companies, coal mining companies, and electricity generators, are top of the list. The value of these assets globally is around $1–2 trillion. That sounds big, but it isn't relative to the total stock of global wealth, which is around $400 trillion. So we are talking about wiping out less than 0.5 per cent of global wealth.[52] This is even less troubling given the concentration of wealth.

CORINNE: That's a pretty powerful hypothesis. Distributional analysis of the green transition tends to miss this point about how concentrated the ownership of stranded assets is, the "assets of destruction". It also tends to focus on incremental effects on the government finances, due to subsidies, or the effect of energy taxes on household disposable income, without taking into account the effects of the broader economic stimulus.

ERIC: One of the ironic consequences of the vast inequality in the ownership of wealth, is that we don't have to worry too much about the losers. The direct and indirect ownership of corporate assets is concentrated in the top 1 per cent.[53] In a situation where we have to wipe out the value of a significant chunk of the capital stock, that cost is actually going to be focused on very few wealthy owners. If you look at stock market ownership globally, roughly 90 per cent of the assets are held by the top 1 per cent of the global population.[54] The value of fossil fuel assets has already been falling as a share of global stock markets, and this may well continue without the overwhelming majority of the population even noticing, particularly as green assets are already taking their place. Putting aside climate change, from a purely economic perspective, this is an opportunity to transform the economy, generate a positive return on capital, and the costs of the transformation are going to be borne by a very small segment of the population.

CORINNE: I would hazard this is counter to popular perception, which is that there are going to be many losers and households will face higher bills. It is critical that leaders and influencers shift the narrative and perception on this in the broader population. Other than in very specific geographies, we don't need to worry too much about the distributional implications of this transition. The owners of capital on average have done extraordinarily well over recent decades and can cope with some losses.

ERIC: It is accelerated creative destruction.

CORINNE: The owners of high street chain stores have suffered terrible losses since the advent of the Internet, and we don't stop and say, "this is terrible, let's bail them out".

ERIC: An awareness that we are primarily creating a huge stock of new assets through investment is critical, and that financing this can strengthen the state's financial position if we want it to. And it needs to be understood that the losers, the owners of the big carbon assets, are a small minority of the globe. We should caveat that this focus is on the distributional consequences of destroying fossil fuel *assets*. There will be other potential costs for households, which is why we also need to use EPICs to prove that most of the population can benefit from tackling climate change.[55] That must be a fundamental political objective. The aggregate population doesn't just benefit from cleaner air, better insulated homes, and knowledge that the safety of the planet is being improved, but there is an economic benefit to the majority of society. We shall explore ideas about how to ensure the returns on this transition are fairly distributed, through a "green national endowment", when we discuss how to supercharge the nation.

CORINNE: Ok, that seems like a good point at which to take stock. At a global level, most experts agree that we are totally off track to halve emissions by 2030, which is a prerequisite for preventing global temperatures rising by more than 1.5°C by 2100.[56] Dieter Helm argues very convincingly that we need a far more aggressive change in course than anything which has been achieved so far, if we are to stand a chance of hitting these targets. He is right, emissions are still increasing, but the task is now much more achievable due to the leaps in technological progress required to transition. Our vision is that accelerating the process is not just realistic and necessary, but is in fact more likely to deliver positive feedback, at multiple levels. This is essential to success. If we can show that the green transition brings quick rewards in terms of lower electricity prices, safer cars, better jobs, cleaner cities, higher standards of living and better lifestyles, we will gather momentum and support to complete the course to net zero. We shall show how to create simple, extreme incentives to alter the actions of companies and individuals. And we shall also describe how, as activist individuals, we can change norms of behaviour and government policy. We have targets, we need actions. We are describing what *should* and *can* be done.

2

Corporate philosophers

ERIC: When it comes to climate change most people think capitalism is part of the problem, that the profit motive pays no regard to the planet. So, Corinne, when did this change?

CORINNE: During lockdown.

ERIC: Your view is that an irreversible culture change is occurring. Is that right, and if so why?

CORINNE: If you had told me five years ago that in 2022 most major financial firms and banks would have signed up to getting their "financed emissions" down to net zero by 2050, that the big oil companies would be transitioning to renewables, that consumer goods companies would have started the hard work of reducing emissions across their supply chains, I would not have believed you.[1] Well, they are. We need to understand why, we need a concrete measure for what they're doing, and we need to work out how to supercharge the process. This culture change means that the private sector is primed to be highly responsive to smart policies. That's the opportunity, we must not miss it.

ERIC: So why has global capitalism suddenly woken up to the reality that it is destroying the planet?

CORINNE: Rebecca Henderson opens her book *Reimagining Capitalism* with a quote from the biologist Edward Wilson: "We have paleolithic emotions, mediaeval institutions, and godlike technology".[2] That is not something we are going to change. So if we are to accelerate this transformation of the economy, we need to shift incentives in a way that is realistic about how human beings operate. This same insight holds for corporations.

ERIC: Are businesses also motivated by paleolithic emotions?

CORINNE: Yes. The corporate version is profit and loss. That's the key operating logic of business – making money.

ERIC: So explain the relationship between responding to the climate crisis and making money. Why has business decided that cutting emissions will boost the bottom line? Isn't the opposite more plausible, that there's a tension between profitable growth and acting on climate?

CORINNE: That framing is too simplistic. Three discrete forces are at work, all of which affect company profits and the incentives of senior executives. Share prices are already being impacted by environmental factors. This is a huge motivator. Stock prices determine companies' ability to access capital for funding their businesses, the cost of making acquisitions, and many executives' remuneration is linked to the stock price. The second fundamental shift is typified by what I call "corporate

Camus", the modern capitalist version of the French existentialist philosopher, Albert Camus. Existentialists believe that the awareness of death gives our life meaning. Mark Carney, former head of the Bank of England, is a "corporate Camus". The fear of bankruptcy gives these companies motivation. Carney doesn't mince his words: "Companies that don't adapt will go bankrupt, without question".[3] Not all business leaders see it quite so starkly, but climate change is increasingly recognized as a source of risk for all businesses. The third motivating factor may be the most potent: do-gooders think they can get rich. There is a boom in environmentally-oriented start-ups with venture capital funding, and many established companies are rapidly building up new environment-facing products and services. Significant areas of the private sector now realize there is big money to be made in the transition to a decarbonized economy.

ERIC: Let's park the idea that share prices are now being affected by carbon intensity, we shall cover this in more detail when we come to discuss money. First, I want to better understand the effects of these corporate philosophers, and the do-gooders who think they can make hay. A corporate existentialist sounds like an oxymoron. Tell me why it isn't.

CORINNE: It is dawning on many business leaders that climate change has opened up a new set of risks. Every sector faces some degree of disruption from climate change itself, what experts call "physical risk". This could include a factory flooding, or workings being exposed to heat stress. Every sector is also exposed to the risk of not keeping pace with how the global economy is responding to the climate crisis. This is called "transition risk". Social stigma, shareholder exclusion, and activist attacks via social media are threats to any business for whom brand and reputation matter. And regulation challenges even the smallest businesses.[4] Many corporations realize they simply won't exist in their current form on a ten-year view. At the extreme, a set of companies face the visible threat of extinction. This is obvious in areas such as oil and gas, or coal. Those that recognized this early, companies such as Denmark's Orsted or BP, have either transformed their businesses already or are in the process of doing so. Of course, some aggressive polluters will try to continue operating under the radar. But the risks they face are rising rapidly.

ERIC: Say more.

CORINNE: In reality, most businesses are highly risk-averse, and the risks associated with climate change are popping up everywhere. Governments are threatening to impose carbon border taxes on imports into Europe, brands are highly vulnerable to activist guerilla tactics, and environmental regulations are hitting most industries.[5] All serious businesses engage in risk analysis, and climate is rising to the top of the list. This results in changing behaviour.

ERIC: What I find convincing about your description of corporate cultural change is that it is grounded in a very realistic set of motivations. Tell us about the do-gooders. Who is getting rich, and how?

CORINNE: Every disruption brings huge opportunities for wealth creation. This case is no different. Many people working in business say they are reminded of the early days of the Internet.

ERIC: Any major technological revolution creates opportunities and new services which are not easily foreseen. It is often easier to identify the areas in decline. Economic history strongly supports the view that whenever major innovation occurs, the net impact on standards of living, growth and incomes is positive despite old industries dying, in this case fossil fuels. The investment-led policies which we are proposing should be expected to create rapid growth in sustainability innovation and investment, which will raise living standards significantly. Nicholas Stern, who is not prone to hyperbole, says, "The experience of previous waves of technological change suggest not only a dynamic period, perhaps a few decades, of innovation, investment, creativity, opportunity and growth, but also large and growing markets for the pioneers."[6]

CORINNE: Quantifying the significance of what Stern describes is very hard, but I am observing this process directly. Within the consumer sector, there is an extraordinary explosion of new categories, such as plant-based dairy, on a scale which no one could really have predicted. Alternative milks are now consumed by close to 40 per cent of US households.[7] Plant-based sales have grown in tandem with the decline of dairy milk. Companies such as Oatly, which listed on the stock market in May 2021 at a value of $10 billion are current winners in this particular battlefield, while incumbents such as Danone are trying to catch up.[8]

ERIC: Are newcomers better positioned to capture this opportunity than existing businesses?

CORINNE: Yes and no. It depends hugely on the sector. In the consumer sector, new brands are disproportionately driving growth. A fresh brand has the advantage of no baggage. It is entirely associated with the new product in question. Although, as we have seen in the auto industry, incumbents can transition too, often very successfully. All the major global car manufacturers have electric vehicle ranges. That is the result of a huge R&D and investment drive over the past five years. In the energy sector, there is the much-lauded example of the Danish company, Orsted. The speed of Orsted's transformation has been extraordinary, as are the returns to its shareholders. Ten years ago it was an oil and gas company. Today it is the largest producer of offshore wind power, and over the past five years its share price has risen over 300 per cent.[9] Over the same period, the share prices of the major oil companies have fallen.[10]

ERIC: This is interesting. It's another reason why the psychologist Dan Ariely is both right and wrong. Humans appear more motivated by short time horizons. Behavioural psychologists describe this as myopia. Investors in stock markets are also described as notoriously short-termist. Yet stock markets can engage in a form of time travel.

CORINNE: How so?

ERIC: Markets get very excited about companies which are growing rapidly, and very pessimistic about companies facing long-term decline. This affects share prices in the short run.

So preoccupations about the long term, or blue-sky dreams about the future, can create short-term losses or rewards. The euphoria around initial public offerings (IPOs), when companies first list on the stock market, is a case in point. As you said, Oatly IPO-ed at a $10 billion valuation. That's equivalent to approximately 20 times its revenues and for a company yet to make a profit.[11] Investors' short-term speculation has the ironic effect of providing cheap money to a business investing with considerable risk for potential profits in the distant future.

CORINNE: These returns, and the publicity around them, provide a huge motivation for entrepreneurs and venture capitalists who want to create and finance businesses which have a positive environmental footprint. They don't just believe it is the right thing to do, they also think they can make money from it.

ERIC: It is undeniable that there is a profound cultural shift happening in the corporate world. It has spread throughout the global private sector in recent years, and is now affecting many companies listed on the stock market. The causes are not fluffy, they are grounded in fundamental motivations. The three forces you cite all impact the bottom line sooner or later. One question that follows is "Does this change in corporate culture matter?"

CORINNE: Until recently, I would often hear corporate leaders ask, "Is sustainability a fad?" A few dinosaurs still think it is, but now that it impacts their profitability, everyone is becoming a convert. This cultural shift matters for the simple reason that the responsiveness of the private sector to smart policy intervention is likely to be high. The private sector is primed to respond to EPICs, smart regulations and targeted activism. But there is a huge challenge in measuring the effects of what companies are doing. Objective measures are required to hold them to account. Otherwise, spin will dominate substance.

ERIC: How can we hold companies to account?

CORINNE: Other than mandated regulation, reporting and target-setting are the two effective ways of creating accountability. Forty countries have mandatory requirements for emissions reporting.[12] More than 9,600 companies voluntarily report detailed information on their emissions to the Carbon Disclosure Project.[13] Transparency is increasingly expected. This has important behavioural consequences. Requiring companies to measure emissions will itself alter behaviour because businesses just like individuals are susceptable to availability bias. Many businesses may be surprised by the sources and magnitudes of their emissions, and peer comparisons reveal the worst offenders. Target-setting is now commonplace, and combined with reporting, it is the way to make companies accountable. The science-based targets initiative, or SBTi, is central here.

ERIC: There's a risk of death by acronym, but can you give us a quick summary of the science-based targets initiative?

CORINNE: The initiative was convened in 2015 by a number of leading NGOs and international agencies with the goal of enabling companies in the private sector to set science-based emission reduction targets.[14] Under this framework, there has been huge uptake by corporates, particularly since 2018.[15]

ERIC: Can you give me an example of these science-based targets, without getting too technical.

CORINNE: Walmart, one of the world's largest retailers, has a target designed in conjunction with the SBTi.[16] Science-based targets invoke the concept of "scopes", an ingenious framework for companies to take responsibility for all the emissions they affect, directly and indirectly.[17] So Walmart, for example, is not just responsible for the emissions created by running its stores or warehouses. It's also responsible for its supply chain.

ERIC: Nachos?

CORINNE: Consider a packet of tortilla chips. There is a relationship between Walmart and the emissions that come from growing the corn, processing the corn into tortilla chips, packaging the tortilla chips into plastic, and transporting them to the warehouse.

ERIC: How much of this does Walmart control?

CORINNE: Most of this happens outside of Walmart's direct control, it lies instead with its suppliers and the original manufacturers. The "scopes" framework makes Walmart responsible for the entire supply chain, which it can influence. Scope 1 captures emissions created directly by the firm. These tend to be low outside of heavy industry, and in Walmart's case are probably not very significant. They might include the fuel burned in a company-owned car, or leaked refrigerants. Scope 2 concerns "purchased emissions", such as the electricity to run offices, warehouses, stores, or air travel of employees. Scope 3 identifies indirect emissions. These tend to be the largest for consumer-facing companies like Walmart, and are usually generated by the supply chain, or by customers. In business jargon these are emissions generated both "upstream" and "downstream" in the value chain.

ERIC: So "upstream" emissions would be those created before the bag of tortilla chips gets to Walmart. What's a "downstream" example?

CORINNE: Downstream emissions are created *after* Walmart sells the tortillas – in this case, emissions created by the landfill when the bag is disposed of, or in the case of a pair of jeans, the energy used to wash them. Downstream emissions are caused by the item after it has been sold to the customer.

ERIC: So the scopes framework makes companies take full responsibility for all the emissions they influence?

CORINNE: Exactly. Without this framework, Walmart would be seen as having trivial emissions. In reality, Walmart facilitates a much broader network of emission-producing suppliers and one that it has the purchasing power to influence. The science-based target framework forces companies to take a systemic view.[18]

ERIC: How widespread is the adoption of these targets?

CORINNE: The SBTi framework has become the gold standard. By the end of 2021, well over 2,100 companies, accounting for more than one fifth of the global stock market, have science-based targets.[19]

ERIC: Isn't there a risk that the current CEO sets an ambitious target for 2030 or 2050, knowing full well it will fall on someone else to deliver?

CORINNE: In order for a target to be approved by the SBTi, the company needs to demonstrate a credible route to delivery. I agree there is still a clear mismatch of time horizons – most CEOs are in place for five years, and are focused on what they will achieve in that timeframe. I think the pressure from financial markets is key to holding CEOs accountable. Risk assessments can affect share prices, and although it may seem technical and unimportant, reporting by companies on climate risks, and the scale of their emissions is a critical component of that pressure.[20]

ERIC: OK, so let me summarize your observations. There's a cultural shift in the global private sector. Climate change is perceived to affect business in three direct ways. It poses an existential threat to some, a huge opportunity for others, and enters risk assessments for all. Large swathes of the private sector seem to be taking these issues very seriously, committing to emissions targets verified by international third parties, and there is now widespread disclosure of climate risk assessments. Despite this shift in corporate beliefs, the reality of this transition is that it will involve near-term costs for many businesses. In order to accelerate the process, and ensure that those businesses which are at the forefront of climate action are advantaged financially, we have to design policies that recognize this reality.

This cultural shift makes the global private sector primed to respond to the right policies. In order to supercharge the process we need to heavily skew the system in favour of innovating new businesses and those making the greatest effort to change in emissions-intensive sectors. In simple terms, the objective of policy should be to tilt the odds in favour of the best corporate citizens and to punish the cheats and free-riders.

CORINNE: Let's talk more specifically about how we can do this with smart taxes and smart regulations.

ERIC: We define "smart" taxes and regulations by two criteria. They must be politically viable and they must succeed in changing behaviour. This contrasts with the usual goal of taxes, which is to optimize the trade-off between efficiency and revenues. We are focused on changing behaviour. There is a role for a carbon tax here, and it could be broadened outside of the worst offending sectors, where most carbon taxes have been directed. But the complexity and unintended consequences of attempts to create an all-encompassing and perfect carbon tax is a problem. It won't be a silver bullet. There are likely far more effective uses of political capital.

CORINNE: My preference is to tailor the solution to the problem, and to be flexible about how we tax and incentivize best practice. An effective way to target corporate behaviour across all sectors of the economy is to introduce a "contingent carbon tax" (CCT) on profits. Can you explain what this is, and how it differs from the standard idea of a carbon tax?

ERIC: Firstly, we would mandate that all companies over a certain size report their emissions using the scopes framework. Governments would then put companies on notice, and warn them that if they do not get close to industry best practice on emissions, they could face a windfall tax on profits within five years.[21] This would incentivize companies to compete on cutting emissions. It also tackles vested interests in a fair and transparent way For example, there is huge variance across electricity utilities in their sustainable ambitions. This is understandable. Many of them have a vested interest in prolonging the life of coal or natural gas assets – because they own them. To see what *can* be done, we need to look at utilities that don't have those assets, and which are investing far more rapidly in alternatives.[22] Merely announcing a contingent corporate tax would create a significant risk to all businesses in every sector. They have to measure their scope 1–3 emissions, and see how they compare with competitors. If they underperform, they have a new tax liability. Nobody wants that.

CORINNE: As the reporting capabilities improve, and they already are, this could be extended to companies of all sizes. The key issue is why should only good corporate citizens, or those in the public eye, report emissions? All companies should. And all companies should be made liable for emissions in excess of best practice. That is a minimum requirement. This seems compelling, so what are the objections?

ERIC: There are some serious challenges, not least the complexity of measuring and comparing emissions. The risk with any tax on company profits is also that accountants try to game the system with complex corporate structures, or fraudulent reporting. We need to be certain that this tax incentivizes rapid change, and not a new industry in clever cheating. The beauty of the *contingent* tax is that we are not actually trying to raise revenue. We would rather all companies begin rapidly reducing emissions. If they all get close to best practice we don't have to impose any windfall taxes. The objective is to encourage companies to change behaviour and to tilt the playing field in favour of good corporate citizens. This tax incentivizes a race to the top. Using the scopes framework, intra-industry comparisons would have to be on a like-for-like basis, so although a given corporation could be liable for the tax, the emissions comparison would be based on the performance of a specific division, rather than on the company at the aggregate.

CORINNE: Otherwise, a mining conglomerate could just sell its coal assets to meet its emissions targets.

ERIC: Exactly. Under our proposal, if that coal asset continues to produce emissions above industry best practice it will be liable for a windfall tax. So changes in corporate structure cannot eliminate the tax liability.

CORINNE: What is also powerful about announcing a contingent tax along these lines is that it can immediately impact corporate behaviour. It sets the process in train and puts the business in control of how much tax it will be liable for in five years' time. A company can ensure it does not pay the tax by sharply reducing its emissions. This seems compelling. Are there other ways we can supercharge companies?

ERIC: Yes. It is not entirely unfair to suggest that economists have been looking at the wrong chapter of the microeconomics textbook. They went straight to the chapter on externalities, and skipped over the earlier one on the price elasticity of demand.

CORINNE: Externalities often frame analysis of effective carbon taxes. This is the idea that businesses and individuals producing emissions should pay for them, and provides an ethical basis for a tax on emissions. The "price elasticity of demand" is economic jargon for how sensitive consumer demand is to a change in price. You're right, it's odd that this is not a central focus in the economics of climate change when we are trying to change behaviour. A key determinant of the sensitivity of demand to price is the existence of substitutes.

ERIC: Exactly. That is why we are repeatedly focused on creating substitutes and targeting their prices relative to that of the emissions-intensive alternative. We need a smart carbon tax framework aimed at *relative* prices.

CORINNE: Ok, how does this differ from a standard carbon tax, and why might it be superior?

ERIC: It's important to be crystal clear that we are not trying to raise government revenue, this is something we shall address separately when we come to discuss supercharging the nation. Our priority is to dramatically change behaviour and cause a collapse in emissions. That may seem obvious, but often carbon taxes falter at this hurdle.[23] The typical carbon tax is aimed at the entity directly responsible for emissions, in your terminology it is a scope 1 tax. Taxing the fuel used in our cars is a classic example. The principle seems compelling, at least from the perspective of fairness. It seems right that people responsible for emissions should pay for them. But this simple theory is very complex in practice, and there are major question marks over efficacy and political feasibility.[24]

CORINNE: You and I have discussed this topic at length. If we look at the history of carbon taxes, there seem to be three distinct motivations, not all of which are consistent. Carbon taxes aim to reduce emissions by making them expensive, they are also used to raise government revenues, and finally they speak to our moral sense that those responsible for pollution should pay for it. These aims often conflict. For example, if behaviour changes dramatically, the government doesn't raise any tax revenue. Using carbon taxes to pay for things is inconsistent with rapid behavioural change. Is there a smarter alternative?

ERIC: We need a *relative price* strategy for taxing goods and services. One of the major problems with carbon taxes occurs when we don't have scalable substitutes. Taxes on fuel are a good example of this. If I have a petrol car and I need to drive to work, my behaviour does not change in response to a tax rise, unless the tax is so punitive that I can't afford to work. That is a bad policy at every level. But if the policy-maker raises the tax on one good, and cuts the tax on a close substitute, there will be a significant response, particularly if the price differential is very wide. Behaviour responds to extreme differentials in *relative* prices of close substitutes.

CORINNE: Talk us through a real-life example.

ERIC: We described earlier how Norway and a number of Chinese cities created a rapid increase in demand for electrical vehicles by making them significantly cheaper than fuel-based alternatives, primarily through tax exemptions. This is a *relative* price strategy. It's important to understand why it works – because there exists a lower carbon substitute which consumers are happy to buy. We need to extend this approach to all relevant industries. The objective is very clear, relative prices must strongly favour the green alternative. How we get there needs to be flexible. Sometimes we should use sales tax exemptions, sometimes corporate taxes. Norway has used a range of exemptions on taxes, parking fees and tolls. Let's consider the steel industry. Steel manufacturing accounts for roughly 8 per cent of global CO_2 emissions, so it is critical to a net zero strategy.[25]

CORINNE: Explain how we can use taxes, EPICs and smart regulations to implement a relative price strategy for steel.

ERIC: The extreme heat required in the manufacturing process of steel is typically generated by burning fossil fuels. The technologies now exist to create hydrogen-based alternatives, which can produce almost carbon neutral steel, although this also requires a sustainable source of electricity to further reduce emissions.[26] One of Europe's largest steel companies is already producing some green steel, as are a number of newer growth companies.[27] Currently, the technologies are not sufficiently profitable at scale. A smart tax needs to change this equation. For example, we could reduce corporate taxes on those companies producing more green steel, exempt green steel from sales tax and import duties, and do the reverse for carbon-intensive steel production. Now this is clearly a *form* of carbon tax, but the critical component is targeting the relative price of the green substitute. To the consumer of steel, there is no perceived difference between steel produced using fossil fuels or hydrogen. They will buy whichever is cheaper. So we need to structure taxes and tax exemptions to ensure that green steel is the most profitable to produce and the cheapest to buy. That is how to create change at speed, and incentivize scaling of new green technologies. The same strategy needs to be applied across all sectors.

CORINNE: Some people will argue that this is what a carbon tax achieves. What makes our proposal so different?

ERIC: Our policy is directly targeted at the relative price of substitutes, and we have a preference for tax exemptions over tax increases. Our policy is also aimed at *scaling* substitutes. Currently, green steel is far more expensive to produce. We need to make a huge effort through tax exemptions and other policies to ensure it is priced competitively. Through time, innovation and economies of scale will make it cheaper, but the initial stimulus needs to be policy-driven.

CORINNE: The change relative to the status quo is also important, something we touched on in the discussion on EPICs. Despite what the microeconomics textbook tells us, there is a striking difference in the perception of consumers, and therefore in the political response, to a tax exemption, as opposed to a tax increase. We are taking this into account. That is the first key point. The second is that the effectiveness of any policy along these lines is almost entirely determined by whether or not close substitutes exist. Policy has to be focused on *relative* prices, and this applies equally to cars, steel or burgers.

This discussion of the steel industry brings us to our final corporate-specific proposal, green trade agreements (GTAs). Before we explain how these might work, we need to clarify the problem they are designed to solve.

ERIC: There are industries, such as steel and cement, where technology exists to produce green substitutes, but vested interests globally can work to block innovation and foster free-riding.[28] For example, what is the point of Europe incentivizing green steel, if imports from China or Russia can simply undercut European production with cheaper emissions-intensive products?

CORINNE: Europe has proposed a carbon border tax to deal with precisely this issue. This is a sensible idea, which Dieter Helm, amongst others, has long advocated.[29] The idea is to incentivize countries which are free-riding to raise their emissions standards. The US and Canada are also considering similar policies. There is, however, a concern that domestic producers will simply see this as an opportunity to further their own agendas, and it becomes a form of protectionism.

ERIC: That's right. Europe's strategy so far has been more astute. Much like our contingent carbon tax, Europe is hoping that the threat of a border tax will itself instigate change in countries which are dragging their feet.

CORINNE: What is lacking is a coherent global process to address these sectoral transitions. Our view is that the carbon border tax should be held in reserve as a stick in order to ensure compliance with hard-hitting GTAs. Explain how this would work.

ERIC: Typically, international trade agreements are made between countries, or regional blocs, and they aim to provide rules covering almost all sectors of the economy. Although environmental standards are increasingly being introduced, our proposal is more direct.[30] Let's continue with the steel industry example. We want the major steel-importing countries to agree on the sectoral standards for global green steel, with an ambitious timeline for 100 per cent green steel. Access to these markets will depend on meeting these standards. Independently, the leading steel-exporting countries should agree to a set of EPICs to incentivize creating the capacity at scale. The World Trade Organization (WTO) could convene and formalize these industry-specific trade agreements.

CORINNE: There is a risk that any multilateral agreement of this sort gets watered down. In contrast, to standard trade agreements, climate change is not amenable to compromise. Europe needs to keep the carbon border tax in its back pocket as a stick.

ERIC: Industry-specific trade agreements would also address important challenges that Adam Tooze and Dieter Helm identify. Tooze points out that the US and Europe are taking very different approaches to changing corporate behaviour, and Europe's reliance on a carbon border tax threatens a green trade war. Dieter Helm has shown how many advanced economies have substituted domestic production for imports in a bid to reduce emissions generated at home.[31]

CORINNE: The advantage of industry-specific green trade agreements is that the major importing economies can aggregate their market power to set rigorous and ambitious standards.[32] And exporters must agree on how to incentivize their industry to deliver. All the relevant parties need to work together, and if one country tries to block progress, there could be grounds for effective sanctions or simply exclusion from the global market.

ERIC: That's right. Unanimity should not be a precondition of agreement. Green steel would not only be exempt from taxes, the relevant countries where most of the world's steel is produced would actually agree to a set of EPICs to supercharge the industry's transition, such as zero interest loans. In this way, we can target the relative price of green steel. The objective is very simply to make green steel far more profitable to produce and cheaper to consume. There should also be targets for green steel as a share of global steel output, looking to phase out emissions-intensive production entirely within the next decade.

CORINNE: This also, of course, requires as much green electricity as possible, because large supplies of electricity are essential to greening steel production. This would create an incentive for economies like Russia and Brazil to transition to sustainable electricity, as it becomes a precondition for their steel industry. We could see a world where the Russian and Brazilian steel industries are lobbying their governments for a rapid transition to renewable electricity.

ERIC: Once onerous standards are set by importing economies, we can get all the world's major steel producing countries – China, India, Japan, USA, Russia, Brazil and Europe – into a room to agree on a common strategy to use EPICs to change the economics of global steel.[33] The importance of this kind of agreement cannot be overstated, the global steel industry's emissions are about eight times that of countries such as the UK.

CORINNE: OK, so let's summarize. We now have global standards for measuring what companies are doing and how to hold them to account. The extraordinary change in corporate culture that has occurred is motivated by three forces. Share prices are now being affected by companies' carbon footprint. At the extremes, some companies fear extinction. An emerging group of rapidly growing businesses, often funded by venture capital, see a profit opportunity. The policy challenge is twofold. First, to maximize the financial equation companies are facing through EPICs and smart taxes. This needs to specifically target the relative price of the green substitute, sector-by-sector. Second, to level the playing field through smart regulations as much as possible, so bad actors don't free-ride and we minimize the collective action risks.

ERIC: So, we have two objectives: accelerate this process of corporate change and incentivize those at the forefront of decarbonization.

CORINNE: Using the framework of EPICs, smart taxes and smart regulations, we have outlined three policy proposals. The first is a smart tax, a contingent carbon tax (CCT) on companies. The beauty of this tax is that we can immediately put companies on notice: if you don't start reducing your emissions, you will accumulate a tax liability. Tax authorities will require all large companies to provide reports on their scope 1, 2, and 3 emissions, with a view to eventually extending this to all companies. Underperformers will be liable for windfall taxes. All the data will be made public. The CCT also mitigates against free-riding. By using the scopes framework, we incentivize large companies to influence their suppliers, and encourage their customers to change too. This has the potential to accelerate and deepen cultural change.

ERIC: Our second proposal combines the idea of positive incentives with smart taxes. We want the government to use a combination of taxes and tax exemptions to target relative prices, where substitutes for a polluting product exist and can be scaled. This is different to the standard carbon tax because it focuses very specifically on relative prices, with a preference for tax exemptions. This approach can be applied, sector-by-sector, to anything from electric vehicles to home heating systems or plant-based burgers. The job of the tax system is to ensure that the sustainable alternative is far cheaper to buy and more profitable to produce. We can't emphasize enough the non-linearity of incentives. The problem with many existing financial incentives in this space is that they are marginal. Human behaviour is not easy to change, only extreme incentives will do.[34]

CORINNE: Finally, we propose global green trade agreements (GTAs) to target climate-strategic industries, such as steel. This fresh approach to trade agreements offers far less scope for vested interests to block progress. We want the world's largest importers to set the standards, and the world's largest producers to agree to a set of EPICs to deliver. Steel is the obvious starting point. Cement should be next. Similar agreements could also be reached for metals, chemicals, plastics, machinery and electronics. Again, the objective is very clear, through a combination of taxes and tax exemptions we want to make green substitutes in emissions-heavy industries more profitable to produce and cheaper to purchase. And again, we want this differential to be extreme.

ERIC: It is finally worth emphasizing that this shift in corporate culture is essential. It amplifies the effects of these smart policies. This is true not only for the policies we have just outlined, but also for our proposals to supercharge the greening of global electricity generation when we come to discuss the nation and the world. But before we get there, we need first to consider the changes occurring in financial markets, the lifeblood of the global economy. So, Corinne, can capitalism go green?

CORINNE: Business has the tools to build the new economy we need. But they aren't going to get there fast enough without the right policies. They need supercharging.

3

Money gets the message

CORINNE: Eric, you have analysed the effects of the environmental wave in financial markets. Can you give us a snapshot of what you see, and why you think it is happening? Then we can get to the crux of the matter. How do we harness these forces for supercharging the green transition?

ERIC: A focus on environmental investing has gripped financial markets, particularly in Europe. Huge changes are occurring, in a way that is very similar to your description of corporate change. There is evidence of environmental factors affecting stock prices and bond yields.[1] You identify this as a motivation for companies to change behaviour. I couldn't agree more. CEOs and boards obsess on their stock price, not least because it affects their personal finances.

CORINNE: Why is this happening now?

ERIC: There are three reasons. Social norms are changing, regulatory changes have been introduced, and stocks perceived to be environmentally friendly have outperformed the market.[2] ESG stands for "Environmental, Social and Governance", and has become the buzzword in finance for investing with a conscience. The trend completely mirrors your description of the change in corporate culture. The shift in beliefs in the past two years has been rapid. It feels similar to the dot-com boom when in the space of a couple of years something that was originally the domain of geeks started to dominate all discourse. The same thing has happened with green investing. Two years ago, clients rarely raised the issue of sustainability within investment portfolios. Now it comes up in every meeting. If we go back ten years, the only serious work in this area was in the area of "impact investing".

CORINNE: Impact investing targets the social return as well as a financial return.[3]

ERIC: That's right. Prior to 2019, investors who were focused on the environmental and social effects of investments were operating in a niche. The big change is in *public* markets. Public markets include the stock market and the bond market, where securities are listed, priced and traded. Outside of the banking system, public markets are by far the largest part of the financial sector. The value of global stock markets exceeds $100 trillion. Bond markets, where the world's largest businesses and governments borrow, are even larger than this. Pension funds, insurance companies and international investors invest trillions of dollars globally in these markets.

CORINNE: This is where the most dramatic change has occurred?

ERIC: That's right.

CORINNE: Ok, so owners are now occupying Wall Street, and they seem to have acquired a conscience. To some extent this reflects a broader societal change. You also mentioned the out-performance of sustainable investments, and the role of policy.

ERIC: Fashions in financial markets are usually fuelled by prices going up a lot. ESG is not a fad, but it would be naive to deny the role of stock prices. In recent years, green stocks gen-erated significantly higher returns than the aggregate stock market.[4] That always attracts attention, and prompts many people to claim that you can invest with a conscience and make a profit. We should come back to this, because it is misleading, but as a force for behavioural change it is an important data point.

CORINNE: Can you explain the role policy has played in this Damascene conversion within the world of money?

ERIC: Perhaps the most tangible catalyst is a regulation introduced in Europe on financial disclosure, requiring that investment funds inform their clients about environmental risks.[5] This fits our *smart* policy description. It is not partisan. Who can object to providing clients with more information about their funds? It is very effective, and it is (relatively) sim-ple. It is a good example of how an innocuous regulation can have a huge effect on behaviour.

CORINNE: Let's pause on this for a minute. Asset managers are now having to disclose their ESG risks, in the same way com-panies have to provide public accounts of their profit and loss. This reminds me of Jeremy Bentham's "Panopticon". Bentham posited the idea of a prison built in an octagon shape around a central courtyard with a watchtower with 360-degree vision to the building around it. Even though the single security guard can't physically be watching all of the inmates at all times, the design of the prison influences behaviour, because the inmates may be being watched at any given moment. Put simply, the possibility of surveillance controls and influences behaviour.

ERIC: That is a great analogy. Although, I think the effects go beyond a reaction to surveillance. Disclosure has profound effects on behaviour. Think about how you would respond. First of all, in order to disclose your ESG risks you actually have to do the work to understand and identify them. That's a significant investment of time and resources, and in the process you may discover lots of previously unidentified risks. So you are suddenly casting your investments in a different light. As soon as publications list all of these exposures, you will be vulnerable to negative press articles and constant questioning from clients. It won't be long before the CEO comes down, taps you on the shoulder, and asks, "Why is your fund causing me such a headache?"

CORINNE: It is striking that a simple disclosure can focus the mind, and can of itself change behaviour. A combination of shifting cultural norms, the superior performance of ESG stocks and regulatory disclosures are changing investor behaviour. What are the effects of all this?

ERIC: One of the central themes in our thesis about how to supercharge the energy transition, is that we need a complete change in the investment behaviour of the private sector. Other than mandating change, this will only happen at scale through financial incentives. Senior management are incentivized by their share prices.

CORINNE: People will be familiar with the advent of shareholder capitalism, which was particularly associated with Reaganomics in the US and Thatcherism in the UK. Widespread privatization programmes in Europe in the 1980s and 1990s, the creation of the single market, and financial market integration accelerated this emphasis on shareholder returns. In the last decade, the shift towards shareholder primacy has spread across much of Asia, most notably to Japan.

ERIC: The net result is that the senior management of most global businesses are motivated by their share prices and the interest rates they pay on their debt. Often senior management are remunerated with shares, and their "cost of capital" affects the profitability of the firm and its ability to create value. This is a huge shift in incentives, which can be a major source of corporate change.

CORINNE: So, companies that are behaving responsibly will see higher share prices, whilst companies that are being irresponsible will face a higher cost of capital. Are there risks from this trend, particularly if it "overshoots" or creates market distortions?

ERIC: My initial take is that these developments are positive because they are causing a change in focus within the corporate sector. These changes are complementary to and reinforcing what you observed advising companies directly. It's all very well saying to mining companies that they need to pollute less, pay their workers more, improve governance, and decommission assets which are polluting. They don't voluntarily pay attention to these concerns, which is why we regulate. But the combination of shareholders relentlessly raising these questions, and markets themselves changing the price of their securities and their cost of debt, is very effective.

CORINNE: It's important to link these changes to our diagnosis, which puts investment spending by the private sector at the heart of the solution. To recap, we need to green global electricity generation, and this requires huge capital expenditure. We then need to electrify transport, buildings temperature regulation, and as much of the industrial and agricultural sectors as possible. We also need vast investment to expand the provision of new products with far lower carbon footprints. We need funding for mini-Musks: technological breakthroughs, for example in air transportation, where we have neither substitutes nor an existing technology to incentivize. How do the changes you are seeing in markets impact this process, and what needs fixing?

ERIC: I think it helps to explain what we mean by the "cost of capital". Apologies if this gets a bit technical, but it is helpful to understanding how financial markets can act as an EPIC. We can then assess where the substantial risks in the process are, and what we need to do to make it work better.

CORINNE: There are two ways for companies to obtain finance. They can raise debt or equity. These two components of the cost of capital – the equity cost of capital and the interest rate on debt – influence companies in very different ways. Equity tends to confuse people more. Can you try and enlighten us?

ERIC: The important consideration with share prices, other than whether they go up or down, is the ratio of the price to the profits of the business. This is known as the price/earnings (PE) ratio. Let's compare two relevant examples. You mentioned Orsted earlier, the world's largest producer of offshore wind power. The stock market currently values Orsted at $70 billion. Orsted makes $3 billion in profits, so the PE multiple of Orsted is 23x. Its shares are valued at 23 times its profits.[6] By contrast, the oil company BP trades on a PE multiple of 10x.[7] You can see why companies are motivated by their PE multiple. If the PE multiple of BP goes from 10x to 23x shareholders will more than double their money, for the same level of profitability. So there are really two ways investors make money owning shares, through either higher profits or higher PE multiples. The PE multiple is effectively the equity cost of capital.[8] A high PE multiple usually signals that the market believes a company is growing rapidly or that it is very high quality and low risk.

CORINNE: So for the same amount of profits, a higher PE multiple means a higher share price. Is there a relationship between companies' ESG scores and their PE multiples?

ERIC: ESG scores are not standardized and are often very misleading. The observations I am making really pertain to the "E" in ESG, and specifically the carbon-intensity of businesses. When it comes to equity markets, it is difficult to provide scientific proof, but I have seen enough to be convinced that there is a significant effect. If you look at the extreme cases, for example companies with high emissions, such as steel, oil, and cement businesses, these are trading on very low PE multiples.[9] The stock market is valuing them less than other companies, per dollar of profit. Like most things in financial markets we cannot assume that this is a stable state of affairs. My best guess is that if the policies we advocate become more commonplace, and the playing field tilts further towards green businesses, these features will become more accentuated.

CORINNE: So there are examples of companies on high PE multiples due to strong environmental credentials, such as Orsted, and companies responsible for high emissions on low multiples. This is an organically created EPIC. Senior management are motivated to have higher PE multiples, and a better ESG score has become a significant factor in that calculation. This is a helpful trend, but is there the risk of a bubble as there was with the dot-com boom? What about companies such as Tesla, which are highly valued by the stock market? At the other end of the spectrum, is there an incentive to sell polluting assets to non-listed companies, who don't care about emissions and are after a fast buck? I'm thinking here of mining businesses with large coal assets. This simply results in asset transfers with no emissions reduction.

ERIC: We need to come back to this. These are live problems. But before we tackle these, it is worth saying something about the cost of debt. Companies mainly finance their activities – such as investing in new technologies, building new facilities or making acquisitions – using the profits they generate, or by issuing bonds.[10] So the interest rate on bonds (known as the "yield") is critical to a business's competitive position.

Unsurprisingly, the cost of debt and availability of credit can be a significant motivator of company behaviour. The evidence suggests that ESG factors are also affecting companies' borrowing costs. Empirically this is easier to pin down, because we can control for other factors. "Green" bonds are priced at lower rates of interest than the bonds of other companies.[11] This is also a potent lever.

CORINNE: These are important trends, but we know financial markets have a tendency to go haywire, and we are interested in harnessing the good, and mitigating the dangers. Financial markets are skewing incentives sharply in favour of "good" businesses and against "bad" ones. That is very helpful for supercharging. The role of policy here is to prevent us from screwing up, and then to harness this potential. So first I want to identify what can go wrong, and where policy-makers need to be focused.

ERIC: There are three challenges, and they all have hyphens: free-riding, green-washing and mis-selling.[12]

CORINNE: Let's take those one by one.

ERIC: Free-riding in this context relates to companies off the radar buying polluting assets from those in the public eye. The publicly-listed companies boost their ESG ratings and the buyer gets an asset on the cheap. The objective should be to decommission or make the assets sustainable, shuffling them between public and private markets defeats the whole process. Consider Rio Tinto, the global mining company. Until 2018 it owned significant coal-mining assets, mainly in Australia.

Rio Tinto is listed in the UK. It is not solely owned by wealthy individuals, but also by some of the world's largest asset managers, who invest our savings through pensions, insurance funds, endowments, and charities. Rio Tinto has embarked on a major programme of selling coal assets to improve its ESG score.[13] If the net effect of shareholder pressure is not to decommission coal production, but to incentivize a fire-sale, then the process is clearly counter-productive.[14]

CORINNE: High emission assets are becoming a burden on publicly-listed companies, so they are willing to sell them cheap just to get rid of them. There will always be buyers for undervalued assets, and they are being transferred to parts of the economy where there is less scrutiny.

ERIC: That's right. It is important to be clear that the private-equity industry, at least in Europe, is also undergoing an ESG overhaul. The major investors in the big global private equity funds tend to be pension funds, insurance companies, endowments and other large investment funds. These investors are themselves starting to impose more exacting ESG standards. This is also happening at speed. Investors are aware of these unintended consequences and are trying to develop more intelligent ways of engaging with companies. Our proposal of a contingent carbon tax would help. If companies sell assets to a bad actor, the new acquirer would be liable to a windfall tax. Of course if the asset is in another jurisdiction, in this instance, Australia, this is very hard to enforce, but there may be ways to address this too.

CORINNE: One challenge is to ensure that no area of finance

should be able to operate under the radar. The brilliant book on commodity traders, *The World for Sale*, written by Javier Blas and Jack Farchy, draws this point out vividly.[15] They point out that more than a quarter of global trade in goods and resources in 2017, more than $17 trillion, was made up of commodities, such as grains, livestock, fossil fuels and metals. Commodities trading is dominated by a very small group of companies, many of which operate with little regulation and make use of low-tax jurisdictions.[16]

ERIC: I worry that ESG investing will drive listed companies, such as Glencore, the world's largest coal trader, to push their fossil fuel assets back into the murky world of privately-run commodity trading. At least if we keep these assets in the listed sector, shareholders can exert pressure for good governance and data is available for public scrutiny.

CORINNE: What should we do to encourage the existing owners to decommission polluting assets responsibly?

ERIC: There are several lines of attack. We can starve bad actors of finance, so they cannot purchase polluting assets. We can try making companies liable for the environmental costs of these businesses, even after they sell them, and we can encourage and incentivize impact funds to buy these assets and decommission them.

CORINNE: Starving bad actors of finance requires the banking system. We shall take this up more directly when we discuss how to supercharge the banking system through monetary policy. Banks can have a huge influence across the entire economic system. If bad actors are excluded from access to credit, they will fail to be material players.

ERIC: Credit is the lifeblood of the global economy, regulatory changes that restrict banks from financing the purchase of fossil fuel assets could be very powerful.

CORINNE: An additional regulatory structure that could be adopted to prevent private companies from gaming ESG criteria would be to borrow the idea of extended producer responsibility, which was pioneered in the European Union over the last 20 years.[17] Under these regulations, companies are held responsible for the product's environmental impact throughout its lifetime, regardless of ownership.[18] This has the effect of shifting recycling responsibility on to producers. Crucially, it also incentivizes producers to think about the design of their product, such as how easy it is to recycle or to make it biodegradable. Could we make existing owners of assets responsible for the environmental effects of them even after they have sold the assets on?

ERIC: This needs to be integrated with the contingent carbon tax. Ideally, this liability would be applied across the life of existing assets owned by businesses. This would also give governments the ability to regulate the global emissions of businesses incorporated within their jurisdiction.

CORINNE: Ok, so to clarify, our proposed contingent carbon tax, which requires companies to disclose their emissions, and threatens a windfall tax on those businesses which are underperforming, could be applied on the basis that companies are liable for the lifetime emissions of assets they own right now.

ERIC: That's right.

CORINNE: There are also impact funds which are specifically acquiring polluting assets and decommissioning them. We can increase the scope and ambition of this sector. Instead of oil and coal assets being sold to owners off stage, why not pressure the current owners to sell these assets at very substantial discounts to impact funds? Some impact funds are specifically not seeking to make a market return, but to maximize the social return.

ERIC: Can more enlightened owners manage decommissioning in a way that protects local communities and infrastructure?

CORINNE: Many of these assets can be effectively repurposed, something the government can incentivize. In many geographies we can maintain the existing infrastructure around coal-fired power stations, and repurpose the land and buildings for solar, wind or other forms of sustainable energy. There are already great examples of this being done successfully.[19] No one anticipates that the Australian coal industry will close overnight. We need to transition this capacity, rapidly, to sustainable sources while protecting the communities and local economies which are dependent on infrastructure and employment. Transferring the ownership of these assets, at steep discounts, to impact funds incentivized by sovereigns, would be far more effective. The Asian Development Bank and the Asian insurer, Prudential, proposed using a mix of "concessional" government or development bank finance and institutional green bonds to buy assets, decommission them, and return investors their expected capital over 10–15 years, rather than the 30 years of remaining operation.[20] This, and other financial models to incentivize decommissioning, need to be scaled up rapidly and copied by others.[21] The discussion on decommissioning often focuses on coal, or fossil fuel assets, but the reality is that there is a much wider set of assets that will need accelerated decommissioning in the coming decades; ranging from petrol cars to home heating systems.

ERIC: Another proposal that would accelerate this process, is to harness the financing power of global sovereign wealth funds for these purposes.

CORINNE: The Norwegian sovereign wealth fund, which was built on oil exports, should use some of its extraordinary financial power to buy dirty assets and decommission them. The Norges Fund has assets of around $1.3 trillion. Around two-thirds of this wealth has been generated by returns on investment and the remainder through selling oil, without paying for the externalities of the CO_2 emissions generated. It seems very reasonable that a significant proportion of this could be devoted to decommissioning stranded assets. The Norges Fund is just one example of concentrated wealth created on the back of fossil fuels.[22] Others would include Middle Eastern and Russian sovereign wealth funds.

ERIC: It is true that oil producers have not paid for the externalities of climate change, and they should bear some liability. But this is equally true of the consumers. I don't know if it is realistic to expect perfect redress from decades past. That said, it seems reasonable that sovereign wealth funds, not just in Norway, but across the Middle East, use a material share of that wealth – say 30 per cent – to offset the emissions that they helped cause. When I go to many developing countries and speak to policy-makers, investors, and businesses, there is considerable resentment about funds such as Norges preaching ESG practices to the rest of the world when that wealth was acquired from the sale of fossil fuels.

CORINNE: Ok, to recap, the first unintended consequence of the ESG wave in financial markets is a form of free-riding, where the listed sector sells polluting assets to owners out of the public eye. This is a serious weakness in the system which needs addressing. We have suggested a series of regulatory and policy responses which could mitigate this practice: restrict creditors from financing the purchase of polluting assets, make businesses liable for the lifetime performance of polluting assets they currently own, and incentivize various impact investors to acquire these assets. Let's turn to the next challenge you have identified in this transformation of the finance sector, greenwashing.

ERIC: "Greenwashing" involves giving a false impression of environmental impact. It can occur with varying degrees of deception, not all intentional. When dramatic change occurs within the financial system, some form of gaming of the system is inevitable. ESG is unlikely to be any different. I want to give a nuanced example. The Polish government, with the help of global investment banks who earn fees from these transactions, was the first sovereign to issue green bonds. "Green bonds" are clearly defined, and the funds raised have to be segregated and earmarked for sustainable investments.[23] Now the Polish government has one of the worst track records in the EU on environmental policy.[24] So we need to stand back and consider what is happening here. One could argue that Poland is simply taking advantage of a captive investment universe and obtaining preferential funding for a part of their budget. ESG funds face a shortage of assets in which to invest, so the Polish government will happily exploit any opportunity provided. But at the same time, I suspect that within Poland there is a battle going on between those with vested interests in the status quo of the coal industry and those trying to promote sustainable alternatives, so at the margin the existence of the green bond market may help.

CORINNE: What can we do about this?

ERIC: Some of this is naivete. The asset management industry, in particular, suffers from insufficient focus on the system-wide effects of ESG investing, and a lack of clarity over the principles governing the process. There is a disproportionate emphasis on measuring ESG and insufficient focus on effect.

CORINNE: Regulators and activists are acutely aware of the risk of greenwashing. This is an area where there have been significant cases of whistle-blowing. One of the largest German investment funds, DWS, has already been investigated by the German financial regulator in response to allegations that the claims it is making on ESG funds are misleading.[25] They will be one of many. These issues can be mitigated by disclosure, higher quality analysis, and a tough regulatory stance.

ERIC: We need to return to your concern that a change in the investment landscape which starts as intelligent, well-intentioned behaviour, may morph into risky speculative activity. There is nothing specific here to ESG investing. Speculative bubbles are when asset prices start hijacking social behaviour. When people who never usually take an interest in markets start talking about ESG investing, that is rarely a good sign for returns. Such behaviour is universal to the psychology of human beings, and some dimensions of "bubbles" can be functional, such as the collapse in the cost of capital for huge amounts of R&D as we saw during the dot-com bubble. To some extent we actually want an ESG bubble.

CORINNE: What about the problem of mis-selling? There is a lot of hype currently that ESG investing will deliver great returns, which seems like a simplification of how markets work. I fear some investors will be disappointed.

ERIC: Agreed. In reality, prospective returns from investing are largely a function of luck, and how assets are currently priced. Inevitably human beings extrapolate transitionary returns into permanent features of the investment landscape, and at some point significant investor disappointment occurs. However, part of this process is highly functional. As a system we are trying to reduce the cost of capital for lower-emissions businesses and impact areas of investment spending, and we are trying to raise the cost of capital on businesses which contribute to the climate crisis.

CORINNE: Is Tesla the poster-child of a green bubble?

ERIC: We don't want to get into the game of calling individual share prices, but the probability that Tesla will generate free cash flow commensurate with its market valuation is low. Of course, Tesla may discover areas of value, such as battery storage, to vindicate its share price, but that's not really the point. The probability of generating a significant return from here is low, and that reflects the current market valuation and the hype around the business. It is the job of professional investors and active fund managers to identify which assets are fundamentally mispriced, and where pricing is reasonable. My own perspective is that concerns about valuations are relatively narrowly-focused. As the corporate sector rapidly transitions towards focusing on ESG, the investment universe will broaden, investors will become more sophisticated, and regulatory reporting requirements will tighten. All of these are forces which should mitigate against both greenwashing and potentially destabilizing mispricing of ESG assets.

CORINNE: This perspective surprises me somewhat. It's not really clear if the emergence of a bubble is good or bad. Can I get a straight answer?

ERIC: I think it is important for securities market regulators to be aware of these trends. As a simple rule, bubbles in the stock market are often bad for investors but good for society.[26] If we get lots of cheap R&D in green technology, fuelled by speculative optimism, that's fine by me. So I think this process is helpful, but R&D is only one part of the solution.

CORINNE: We have discussed the effect financial markets are having on the private sector, which is broadly positive, and has the potential to amplify effective policies. What about an area I know you have been working on, deriving ESG scores for sovereign governments. Can you explain what is happening in this area and how it can help?

ERIC: This process is in its infancy, but progressing rapidly. In the same way that we can score a company based on their emissions, we can score countries. Now that investors are doing this, it could start to affect the cost of debt for governments. If it did, the impact on their behaviour could be quite profound. At a personal level, I have already seen evidence of countries refining or developing their climate strategies in part due to engagement with international investors. As in the case of the Polish example, whatever politicians' beliefs about climate change, they care about their borrowing costs, because it can determine their fate.

CORINNE: So in theory, you could end up with a lower cost of debt for a country with a 2030 net zero target, like Finland, and a higher interest rate on the debt of a country with no net zero target at all. Do you have examples of where this has already been done to effect?

ERIC: Not yet, and as someone who invests in sovereign bonds, I would expect other factors to dominate. But this area is fascinating. After all, even in the Polish examples, the government has an incentive to invest in sustainable alternatives. All governments care about the cost of debt, regardless of ideological colour and regardless of their system of government, because it determines their political possibilities. Incentives are powerful. If waves of global capital re-price sovereign debt markets based on ESG scores, there is a very high probability we see changes in governance. I wouldn't count on this happening now, but it is a real possibility in the future.

CORINNE: The obvious concern here is what gives global capital the right to dictate the policies of sovereign nations?

ERIC: Climate change may be a legitimate exception. The nation state isn't sovereign when it comes to emissions. We need global standards. In this instance, capital could act as a check-and-balance on rogue states. If we want to further democratize this process, there is also scope for consumer and investor advocates to turn the spotlight on all of the relevant parties.

CORINNE: So is shareholder capitalism morphing into a form of green capitalism?

ERIC: More so than I would ever have thought. The ESG wave in Europe looks irreversible. It will have significant effects on corporate behaviour, and on green innovation. In contrast to the advent of *shareholder* capitalism, Europe appears further advanced than the US. But investor capital is global, so the effects are being felt in the United States too, despite resistance. The emphasis on shareholder primacy is far more ingrained in American culture. This is something both you and I have witnessed directly in our day jobs.

CORINNE: I think this process is global, but with significant geographic nuance. The interesting fact is that there appears to be a critical mass of investors who are actually affecting the prices of securities. Despite the challenges that we have high-lighted, I do think it is important to stress that this is huge progress relative to the status quo. Whatever the limitations of green capitalism, it is an improvement on the unfettered version. Financial incentives are being skewed towards better businesses. Speculative growth capital is widely available for green technologies. This has two positive implications which really matter from the perspective of supercharging: the effects of smart policies will be amplified by the private sector's response, and "mini-Musks" are more likely to emerge with time. If the global banking system can be incentivized to re-price and reallocate credit towards sustainable businesses, powerful forces will be at work.

ERIC: As with all rapid system transitions, there are risks. In this instance, it is the spectre of the three hyphens: free-riding, green-washing, and mis-selling. All could yield unhelpful unintended consequences. We need policies to proactively manage these risks. We need to extend the idea of extended producer responsibility to asset-ownership, and we need to incentivize impact capital to decommission emissions-intensive assets and to protect the communities which depend upon them.

4

Supercharge the individual

CORINNE: The more you think about climate change, the more you start thinking that you need to change everything in your life. Even when I'm on my smartphone, I can't help thinking: How much energy was used to make this? How much fuel was consumed to mine the metals, or transport the components from China, California, Malaysia, Taiwan? How much energy is required to heat the Apple store? How much electricity is used to charge the battery? It is hardly surprising that many climate activists conclude that everything in our behaviour, habits and lifestyles needs to change. This also makes the challenge feel insurmountable. At the other extreme, the Canadian climate economist, Mark Jaccard, has written that the focus on the need for individual lifestyle change is a myth.[1]

ERIC: Individual lifestyle decisions are often a central element of "solutions" in climate campaigns. This approach often reads more like a wish-list than a focused strategy for social change. The real challenge is delivering a huge global investment drive to make electricity sustainable and electrify all energy use. The implicit view of psychology is also naive. Changing behaviour is not as simple as responding to ethical arguments and lists in books. And yet, as we mentioned in Chapter 1, there are some important areas where we may not be able to rely on either investment or new technology, and we do need to change behaviour. In which case we need a realistic understanding of social change and behavioural psychology.

CORINNE: The typical to-do list says "eat less meat", "car share" and "consume less". On the fuzzy end of the spectrum, we're advised to practice mindfulness, plant trees, and work to find our purpose. There's something fundamentally unhelpful about much of this, which is worth being blunt about. Even the engaged, informed and the motivated don't change their behaviours like this. I have a guilty history of susceptibility to fast fashion. I have to fight the thrill of filling a digital basket with cheap fashion goodies, the big bag arriving, and new outfits galore. You get a serotonin hit from it. Marketing strategies play on the precise psychological traits that Dan Ariely describes. I've managed to quit fast fashion now, but I still struggle to wean myself off the odd burger. The recommendations in books like *How To Avoid A Climate Disaster*, or *The Future We Choose*, are spot on.[2] As individuals, we should reduce meat consumption and stop flying, at least until green air travel is established. It would help reduce emissions if everyone limited consumption of *things*: clothes, electronics, plastics, household goods or packaging. The point is not that these behavioural shifts don't help, of course they would. But recommending them in books as individual to-do lists doesn't make much difference.

ERIC: I would be surprised if anyone has turned vegan after reading Bill Gates's book. But that isn't a cause for defeatism. The reality is that millions of people *are* becoming vegans, because social norms are shifting.[3] There is a community creating change. We need to understand how this is happening, and then supercharge it.

CORINNE: Before we outline a more powerful approach to thinking about behavioural change and empowering the individual, can you briefly explain why we can't just leave it to technology.

ERIC: There are two reasons. Just because a technology exists, does not mean it will be deployed.[4] Most of the time, the progress we are making actually involves adopting and using technology that has existed for some time. Central to our argument is that something like 75 per cent of emissions can be reduced by huge investment using technologies that already exist. Our main focus is to accelerate the adoption of these clean technologies. This leaves the remaining 25 per cent, where technological solutions are at best a work-in-progress.

CORINNE: A major challenge is food. There are helpful technological innovations, but we need to change norms and habits. This is hard. The overwhelming majority of people really don't care if a car is electric or petrol-fuelled, or if their home is heated by clean electricity or gas. These investments are all about EPICs. I don't really count this as "behavioural change". But shifting to plant-based burgers or eliminating meat and dairy from our diet is a qualitatively different form of change. How do we make the global food industry sustainable? Agriculture currently accounts for approximately 15 per cent of global emissions, and this share is likely to rise.[5] We know that as incomes rise globally, and nations transition from low- to middle- or high-income, people will eat more meat and the overall carbon intensity of the global average diet rises.

ERIC: Emissions are generated at all points along the food supply chain – production, storage, transport, processing, retail and cooking. Some of these will improve with sustainable electrification, but not enough. Which aspects of global food production are priorities?

CORINNE: On some estimates, livestock accounts for around half of total food-related emissions. The role of land is equally important. Our dominant methods of growing food destroys carbon sinks in forests and soils. As consumers, we do need to stigmatize meat consumption, and as activists we need to pressure the industry to produce sustainably. There are also policy levers we can deploy, which we shall come back to.

ERIC: Another major area where the tech solution is not imminent, is air travel. Currently, air travel accounts for a relatively modest 2.5 per cent of global CO_2 emissions, so it is less significant than food.[6] But its share is rising rapidly. In five or ten-years' time, without changes in behaviour, its share of emissions may be far higher.[7] There is significant hope for green aviation on a 10–15-year view, but this is not fast enough to halve emissions by 2030.

CORINNE: This is an area where a cap and trade system makes sense, as a solution to restrict air travel in the least disruptive way. This would require a national or international agreement to fix a cap on business tickets annually, steeply declining each year. Businesses could trade blocks of tickets between them.

The pandemic experience suggests economic activity would not suffer because of less business travel. If businesses really need to fly, they should pay for it.

The third important sector where technology does not yet offer scalable and complete solutions, is plastics. Annual emissions from worldwide plastic production and their contribution to landfills amounted to almost 2 per cent of global emissions, and of course, there are devastating polluting effects of both plastics and microplastics.[8] There are technologies on the horizon that may be able to decarbonize plastic production, but these are nascent, and emissions from landfills remain a huge challenge. On some estimates, reuse and recycling can drive a maximum plastic emissions reduction potential of 60 per cent.[9] This all points to the necessity of plastic demand reduction through a change in consumer behaviour, such as shunning single-use plastic, or opting for alternative products, although this is complex too.[10]

ERIC: Ok, so currently there are three areas where the best strategy to tackle emissions is in fact to reduce demand – eating meat and dairy, flying and plastics. Unfortunately, we see rapid growth in all three when incomes rise. We want average incomes to rise, that's evidently a good thing for most of the planet. But we need to "decouple" higher incomes from more emissions.

CORINNE: These are all areas where incentives may be used to help, and regulations have an important role to play, particularly in the case of plastics. But we also have to think about deeper factors which change behaviour.

ERIC: So let's start with a realistic psychology of behavioural change. This is something organizations like Extinction Rebellion and activists such as Greta Thunberg have given a lot of thought, and is often under-appreciated.[11] In the aggregate, human behaviour is dominated by habits, norms, enforced regulations, and incentives. As we have said, most people simply don't change their behaviour in response to a well-made ethical argument. People *want* to do the right thing, but they are most likely to do what others in their reference group are doing, and whatever is convenient.

CORINNE: The first important observation is that there are two different types of individuals. Those who are highly motivated, almost always a minority, and those whose behaviour will only really change in response to the social context. There are leaders and followers. If we are serious about accelerating change, we have to deploy different strategies to influence these two groups. The good news is that the motivated minority have disproportionate power. This is not because of status, but because they are motivated, often highly motivated, and there is an availability bias in our perception of prevailing norms.[12]

ERIC: Say more about what that means?

CORINNE: Our perception of what is "normal", "typical", "acceptable behaviour", or what "most people are doing" is not based on objective statistics. It is biased by small samples which are available. The behaviour of vocal, prominent, small groups can completely alter our perceptions of what we deem normal. Whatever the precise reason is, the evidence is quite compelling that radical change is typically instigated by small organized groups. Having the weight of moral and scientific evidence on their side is also essential. Erica Chenoweth and Maria Stephan document and quantify the history of civil resistance, and conclude that mobilizing 3.5 per cent of the population in non-violent resistance is the threshold for societal changes.[13] This is the compelling strategy of Extinction Rebellion.

ERIC: To help the motivated minority, we need to outline a strategy to amplify the power of activism. There are three areas where we think this can be done: empowering employees, cancelling corporations, and network alliances. These are not comprehensive, but they're a good place to start.

CORINNE: The second type of individual, the vast majority of us, falls into the category of "sheep". Evidence from social psychology overwhelmingly supports the view, as we have said, that most of us only change our behaviour in response to our perceptions of norms, convenience, and short-term reward or punishment. An activist trying to change group behaviour should be focused, therefore, on how to quickly change social norms, cultural practices, and the law. Most people's behaviour changes in response to the context they see around them. They respond to their perceptions of what the group deems acceptable behaviour. The policies we advocate within finance, the corporate sector, national government policy and global policy will reinforce this change in context, but we need activists at every level. And the less tractable problems, which fall outside of EPICs or technological fixes, may ultimately depend on social stigma and a profound shift in collective values.

ERIC: You and I know and work with highly effective activists. Many would not identify with the term "activist", but they have been at the forefront of changing norms and beliefs. They work inside companies, think tanks, government, research institutes, academia, and organizations like Greenpeace and Extinction Rebellion. These people are phenomenal and have worked relentlessly, over decades, to shift our beliefs and collective focus. They have inspired us both. They will find many of the policies we advocate helpful, and although much of our analysis is familiar to them, the framing and strategy, particularly around EPICs, the winners and losers, and how to think about the state's balance sheet, may be useful. These networks of motivated minorities don't really need advice on strategy, they already punch well above their numerical weight. So I want us to focus on a sphere of influence where we think there is huge potential, which is currently under-exploited: employers. As employees we can have a disproportionate influence on our employers.

CORINNE: The last few years has seen employees use petitions, walk-outs or strikes to call out their employers' policy on a number of topics, be it contract worker welfare, sexual harassment, investment policies or climate action. Employees have demanded more ambitious, or coherent climate policies at big name brands such as Amazon, Google and McKinsey. In response to these petitions, I have seen senior executives eye-roll: "they should just quit if they don't like it". But I have also seen board members sweating in terror at the prospect of a petition being leaked.[14]

ERIC: What tactics and objectives would you advise to employees trying to influence their own firms?

CORINNE: I should caveat this by saying it depends where you work. If you are one of only several employees, or perhaps have very little employment security, it may make more sense to join a group unrelated to your work. In larger firms, don't underestimate your power, particularly if you can find a small group of like-minded colleagues. Change always starts with a small group of motivated employees articulating concrete expectations or actions they think their company should be taking. Obama's advice to the Black Lives Matter movement is pertinent: make sure your demands are as specific as possible.[15] By no means assume that there will be resistance. Smart management and business owners should embrace ideas for sustainable growth. But having a concrete proposition makes it much easier to assess if there is a true response from executives, or mere lip service.

ERIC: In your experience, how do employers typically respond?

CORINNE: It is very hard to generalize, but climate change is an issue that cuts through most of the cultural grouping or barriers that can often hinder cooperation across organizations. Time and again, I hear from corporate leaders that employee motivation, talent attraction and retention are important motivators for them to act on climate. They are feeling the pressure. Interviewees challenging their interviewer what they are doing on climate action works. Employees are making it known that they care, and business leaders have to heed this to remain competitive. I see this anecdotally, and it is supported by evidence.[16] Project Drawdown has launched a platform to empower employees to influence their employers on climate, stating "every job must be a climate job".[17] It's an area that is increasingly getting recognition as an axis for disproportionate influence for the average person.

ERIC: What type of concrete issues would you suggest employees can target effectively within their firms?

CORINNE: Behavioural change is very hard. Human beings often know what the right course of action is, and still don't do it. For example, obesity is a complex systems problem like climate change.[18] We have good reason to believe that obesity will not be reversed until we address two features of human decision-making: availability and social norms. As long as there are cheap and widely-available high-sugar processed foods, we will continue to consume them in excess. If all my friends are buying their children iphones, I am more likely to do the same. Human beings rarely form preferences independently of others. We are creatures of what's easy, what's around us, what society condones, and what people we know are doing.

ERIC: In some ways, the goals of our employee activist are similar to that of the policy-maker. We need to push our employers to reshape the choices available to customers, to make them more sustainable, convenient and price competitive.

CORINNE: "Availability" in the climate debate is about how corporations and policy-makers shape the option set for other businesses and consumers. Are supermarkets actively promoting plant-based alternatives? Are automotive companies committing to making a zero-carbon car, so that we have the option of buying such a car? Are commercial real estate companies systematically upgrading the energy efficiency of their portfolios? Are banks using their lending power responsibly? These are considerations that an individual can't directly control, but activist employees, vocal groups of consumers, not-for-profit activists – if they all act in concert – can change the choices available to everyone participating in the economy.

ERIC: Actions companies should be taking will vary significantly by industry. Is the scopes framework we've discussed useful here?

CORINNE: Absolutely. In some business areas, say agriculture, consumer goods, cement or homebuilding, making green products and services the default option for customers might be a central objective for an employee activist. In other sectors, such as technology, business services or financial services, the potential for impact is in Scope 3: the emissions of customers. This could include, for example, tech majors or banks excluding fossil fuel majors from their customer base. There is some debate about the potential impact of this as they will likely have an alternative. Indeed, stopping lending to fossil fuel majors may be counterproductive, given the capital requirements of investing in their green energy businesses. But it does act as strong stigmatization, which drives the overall culture change.

ERIC: Can we also use regulations and taxes to expand choice and strengthen the hand of activists?

CORINNE: The case for advocating a change in product mix and focus at any company will be heavily influenced by the likely impact on profit margins. Raising taxes on food is highly regressive: those on lowest incomes end up paying a greater percentage of their income. If it's done, it has to be combined with measures to protect the most vulnerable in society, such as food stamps focused on nutritious food groups.[19] EPICs aimed at the price of the sustainable alternatives, to make them priced more attractively would be ideal, in combination with taxes where appropriate. In many countries, we already exempt certain food groups and products from sales tax, this principle needs to be applied to all sustainable purchasing options.

ERIC: What about regulations?

CORINNE: It depends where you are. One of the recurrent themes of our discussion is to have singular focus on rapid emissions reductions, but have lots of flexible policy options centred on this goal. Tax exemptions can be an important pathway around ingrained political resistance to regulation, which is prevalent, for example, in some US states. Legislating a requirement to provide, for example, plant-based meat alternatives at all large-scale fast-food franchises is entirely defendable. Both consumers and businesses benefit if the plant-based alternative is sales-tax exempt. That is an example of smart policy-making. If we can align employee activism with these kinds of EPICs, change will occur rapidly.

Beyond our capacity to influence groups of people as an employee, a friend, a family member, or a community participant, we all have the internet at our fingertips. Social media is a potent tool for activism.

ERIC: Social media also has many negative connotations. There are plausible arguments that it hinders many aspects of our democratic process, with the spread of fake news, the manipulative targeting of marginal voters in elections, and the tendency to present the world through an echo chamber. But there is one upside. In certain domains, the world has been substantially flattened with respect to an individual's voice, and the power of activists has been amplified.[20] Also, and this is a unique feature of climate activism, there is no serious vocal opposition. There are of course vested interests, but there is no seriously motivated large group of individuals arguing against action on climate change. Let's talk through some examples, and describe how social media campaigns can work and how empowering it can be.

CORINNE: There are hundreds of examples of how real changes have been made by small groups of activists. The beauty of social media is that anyone can investigate a brand. Go on Twitter, Facebook or Instagram, ask them what their sustainability policy is, ask them about their emissions, where are they sourcing from, what evidence do they provide. If they don't provide answers, call them out. If they are publicly-listed companies, which most big brands are, find out who the shareholders are, and call *them* out. This process is incredibly powerful. You can do it as an individual, or you can join a group.

ERIC: When it comes to cancelling corporate culture, minorities rule. Which are your favourite examples?

CORINNE: I shall briefly discuss three cases of activism, which span different industries and geographies, and illustrate how big an impact ten or 20 individuals can have. One involves Timberland and Nike, and how Greenpeace radically altered their leather sourcing policies in Brazil. This was done through social media. The second shows how the legal process, another vehicle for smart activism, was used to turn the spotlight on Shell's emissions targets. And the third example is how a small minority of shareholders called out the oil giant, Exxon.

ERIC: One thing I really like about the Timberland story is that it actually ended up with Greenpeace releasing a statement praising Timberland's commitment to source sustainable materials in Brazil. It also reveals how effective straightforward activist pressure can be. Although Greenpeace had done a lot of serious research into the sourcing of materials in Brazil, and the impact on the rainforests, the eventual trigger for action was simple pressure via email.

CORINNE: That's right. Greenpeace alleged that Timberland was sourcing leather from Brazilian cattle farmers, who were illegally felling Amazon rain forests to create pastures. The story is well documented by the CEO of Timberland, Jeff Swartz. He recounts the story in a *Harvard Business Review* article, disingenuously titled "How I stood up to 65,000 activists".[21] When you read the full text, Timberland in fact contacted Greenpeace after receipt of the first email. And despite Swartz's protestations, Nike, Timberland and Walmart were all pretty clueless about their supply-chains in Brazil. The pressure from activists got them to focus.

ERIC: Talk me through the Shell ruling, because this is really extraordinary, and as a model it could be replicated elsewhere.

We have already discussed externalities, when the businesses causing climate change are not paying for the damage. In this instance, Friends of the Earth in the Netherlands sought a legal ruling to make Shell responsible for its emissions.

CORINNE: That's right. The Dutch arm of Friends of the Earth, Milieudefensie, sued Shell for violating human rights. A Dutch legal ruling in 2019 had already set a precedent for deeming climate change a real threat to human rights.[22] The district court in the Hague ruled in favour of Milieudefensie, mandating that Shell target a 45 per cent reduction in emissions by 2030. This is far more ambitious than Shell's own target.[23] This sets a precedent and is an objective measure of how much more ambitious our targets need to be. It is worth being aware that currently Shell plans to appeal the ruling, but it demonstrates the legal risk hanging over businesses with large-scale emissions.[24] As activists, we can also bring existing laws to bear to protect our collective well-being.

ERIC: There is an important strategic point that we should clarify here. Someone might reflect on this and ask, "why are you bothering with EPICs, smart regulations, ESG, and repurposing monetary and fiscal policy, why don't we just take all the emissions-intensive businesses like Shell to court?" How would you respond?

CORINNE: To succeed, we have to fight on all fronts. Lawyers should use the legal system, policy-makers should use EPICs, smart taxes, and sectoral green trade agreements. Individuals should use targeted activism. Financial markets have embraced ESG, but nevertheless need regulating and close surveillance.

None of these are mutually exclusive, and when they all happen simultaneously the probability of success rises, not least because norms change.

ERIC: Tell us what happened with Exxon.

CORINNE: This is a great example of a minority of shareholders instigating change. In America, this is often grounded in the argument that climate action benefits shareholders in the long-run, so it is less a debate about shareholder value than time horizon. Tactically, this is astute.

ERIC: Can you explain exactly what has happened?

CORINNE: A minority shareholder, called Engine No.1, an impact fund set up by a veteran hedge fund manager, succeeded in ousting a number of directors on the board and forcing Exxon to develop a plan around carbon emissions. Its objective was to put key directors with experience in energy transition on to the board and to pressure Exxon to develop an emissions reduction strategy.[25] To put this in context, Engine No.1 is tiny in the context of other hedge funds and investors in Exxon. It was only in existence for about six months, owned a fraction of 1 per cent of the shares, and had total assets of around $250 million. For context, Exxon has a market value of around $250 billion.[26] Much larger shareholders, such as Blackrock, and the British investment firm, Legal and General, are believed to have voted with Engine No.1. Now it is important to say that most serious climate activists would still view Exxon as a very poor corporate citizen, but activist pressure can still work.

ERIC: We have argued that individuals can cancel corporate culture, become activists, and amplify their voice through social media. We see the process of "minority rule" in the influence of junior employees grouping together, but the same principle applies at the top. If a small number of CEOs start setting more ambitious decarbonization strategies, others will follow. This will influence the direction of policy, which reinforces these strategies. Citizens are employees, and voters also work for governments. There is a groundswell of forces all pushing in the same direction. It all connects. Our central observation is that small minority groups, often at the periphery of a network can have disproportionate influence in changing accepted norms and ways of behaving. Recommendations focused directly on individual behaviour are not effective. We end up underestimating our own power. The "eat less meat" strategy isn't strategic enough.

CORINNE: Most of the insights we have discussed are not new to social psychology. Our observations about minorities instigating change, with the majority adapting to perceived new norms, are well grounded both in evidence from history and behavioural theory. We have long been aware that motivated minorities are significant drivers of social change, and newer research suggests that modern technology has increased their influence.[27] Before we draw this part of our discussion to a close, I want to consider an aspect of our argument that may unsettle some readers. Your original area of academic specialization was actually moral philosophy. How do you feel about the kind of "minority rule" we are describing?

ERIC: We have referred to the majority of society as "sheep", which is pretty offensive, even if we include ourselves, which we do. We are both very influenced by peer reference points and social norms. That is partly what it means to be human. The policies we are advocating involve a reorientation of behaviour and norms. Many people will view our recommendations as self-evidently ethical, because we face a climate emergency. This is correct. But there are always legitimate questions to be asked about the means. I would argue that human society has always progressed through the ideas, actions and motivations of minorities. It cannot be a coincidence that the majority ends up agreeing with them. Our ethical frameworks appear to converge over time. So much so, that all nations, despite cultural differences and varying interests, can sign up to the UN Sustainable Development Goals. The trajectory of our moral history has a logic to it, such that, to some extent we all come to agree with the minority. We are just behind the curve.

CORINNE: In the sense that I believe vegetarians are right, but I still fall off the wagon?

ERIC: Yes. Recognizing the validity of an ethical principle does not mean adhering to it. Otherwise we wouldn't need enforcement. Humans consistently recognize that what they actually do is frequently in conflict with what they *should* do. We expect enforcement authorities to uphold what is right. A basic principle of state intervention is that regulation of our individual behaviour is legitimate when it affects someone else. This principle is recognized universally in all social organization, otherwise we would have permanent conflict. Climate externalities are real. What I eat affects other people, so I am afraid it's not just up to me.[28]

CORINNE: We started by describing this process as "minority rule", but perhaps climate activists are best described as good shepherds.

ERIC: And being a good shepherd has never been easier or more powerful. If you want to amplify your effect on corporate or political behaviour make alliances with other motivated minorities. If you're an individual or a small not-for-profit, you can also alert the big guns. You can tag Greta Thunberg on Twitter, or Kate Raworth, or Extinction Rebellion, or Friends of the Earth, or us. That's all it takes now for anyone to act as an activist. Any organization in the public eye is sensitive to being called out, but if it comes from multiple sources, it quickly feels like a crisis. So don't stress about using too much water to clean your dishes, get on Twitter and join the motivated minority.

5

Supercharge the nation

ERIC: Let's summarize the argument so far. We have discussed the rapid cultural changes in global business, and the ESG wave in financial markets. Europe appears to be advancing most rapidly in both these areas, but the effects are being felt globally, and are spreading. Politics is also catching up. China, under Premier Xi, the United States, under President Biden, and India, under Prime Minister Modi, are also shifting rapidly.[1] There is a consensus that we all need to move faster. We have outlined a set of principles for smart policy-making to accelerate the transition.

CORINNE: Yes, we have shown that the best policies are *simple, effective* and *non-partisan*. We have provided a framework for categorizing the challenges specific to climate change: *simple maths, mini-Musks* and *herding sheep*. We have also provided concrete proposals, such as contingent carbon taxes, EPICs aimed at the relative price of green substitutes, green sectoral trade agreements, all of which aim to tilt the playing field in favour of sustainable businesses that are driving the decarbonization of our economy. We have explained how to mitigate free-riding and greenwashing in financial markets. We concluded from these discussions that the effects of policy interventions will be amplified because financial markets and global business are primed to respond to EPICs and regulations.

ERIC: In line with expert consensus, our diagnosis puts sustainable electricity at the heart of the global strategy to collapse carbon emissions. A rapid acceleration of investment in wind, solar and other alternatives, as well as infrastructure for energy storage and transmission is needed to get us close to 1.5°C.[2]

CORINNE: The evidence suggests that this is substantially a "simple maths" problem, an economic challenge. In other words, we have the technology and the physical resources, the challenge is to create an extreme incentive structure that rapidly changes behaviour. If the cost of capital can be reduced, and a smart regulatory framework put in place, halving emissions by 2030 is within reach. Although the private sector is best placed to deliver these new industries and investments, the EPIC levers we describe are almost entirely in the hands of the nation state.

ERIC: Bill Gates apologises for using the word "policy" in his book. We maintain that any climate strategy that does not put national policies at the heart of its decarbonization objectives will fail. So, how do we supercharge the nation?

CORINNE: We start with a different apology. The following discussion is all about interest rates.

ERIC: There is a good reason we're spending so long on interest rates. All books on climate change should discuss them, and yet none do. All discussions on the green transition end with the same question: "Who is going to pay for this?" Central to our argument is that governments can pay for it, profitably – because we happen to be in an era of historically low interest rates. We shall outline a number of EPICs which can use low interest rates to mobilize global capital towards the climate crisis.

CORINNE: This discussion might get very nerdy. In order to explain how governments can unleash some epic EPICs we shall need to talk about monetary policy, fiscal policy, bank regulations, balance sheets, and return on capital. Nectar if you're a policy wonk, but if you're not, there's still an upside: you'll get to learn a lot about macroeconomics.

CHEAT SHEET I

Q: *Government debt is at its highest since the Second World War. How can we fund all this investment in green electricity?* **A:** What is rarely mentioned is that interest rates on government debt are at their lowest level in history. In fact, they are so low that most rich countries' interest payments relative to the size of their economies are close to the lowest levels for decades. That's the main reason why we can borrow to fund investment.

ERIC: Before we dive into the details, we should start with some important qualifications. All countries vary, and policies need to be tailored. In our discussion we shall focus on nations where interest rates are extremely low, it will become clear why. This feature of an economy is more relevant to EPIC policy options than geography or income per capita. When we come to discuss supercharging the world, we shall also outline policies that are directly relevant to economies with higher interest rates. Fortunately, the economies where official interest rates are very low account for around 70 per cent of emissions.[3]

CORINNE: Let's explain why interest rates matter, and how they relate to "simple maths" problems. When investors look at any large-scale capital investment, they make a simple calculation: does the return on investment exceed the cost of finance. Investments in long-term physical infrastructure, such as a solar farm, wind turbines, or storage, are mainly financed with debt. This is because the returns are relatively predictable. For example, the returns to a solar power investment are based on the current price of solar panels, electricity prices now and in the future, and how much sun there is likely to be. Once these estimates are made, the revenues which will accrue to the owner of the solar farm are reasonably clear. Now, if this investment is financed with debt, the interest rate on that debt becomes a key determinant of whether or not the project happens.

ERIC: This makes our current circumstances uniquely opportunistic. We have the lowest level of interest rates across the globe since records began. Many governments in the developed world can borrow at close to zero interest rates, or even negative interest rates, for up to 30 years.[4]

CORINNE: And these are fixed rates of interest. So if our governments borrow for ten years to finance offshore wind, solar and grid infrastructure, the economics are not affected because the rate of interest is fixed upfront.

ERIC: That's right.

CORINNE: Harnessing the government's low cost of debt is central to our thesis for supercharging. This enables a much lower "threshold" for projects to be viable.[5] That makes sense. So why is this not already happening?

ERIC: There has been a conflict between the desire to introduce "market risk" into electricity provision, and an intellectual failure to make the case for harnessing the state's balance sheet – its ability to borrow – in an emergency. Regulators need to make investments in the electricity sector less risky.[6] In return for de-risking, we need regulators and the private sector to work on much more ambitious investment timelines. When the government awards a contract in sustainable energy infrastructure, it should essentially guarantee the debt in return for much more ambitious rates of investment.

CORINNE: This is a perfect opportunity for "simple maths" supercharging.

ERIC: The Dutch example, whereby the private sector is exposed to instability of electricity prices, reveals the extent to which the real failure is one of thought and a holistic perspective.[7] The government and public recognize the need for wind energy, and then someone within the policy-making process decides that the private sector should "take more market risk", which sounds sensible on the face of it, but actually misses the point. In an emergency you don't take unnecessary risks. We need to de-risk electricity generation and accelerate investment.

CORINNE: So the challenge is to enable the private sector to borrow at scale at interest rates close to zero, in return for which there is a reduction in their return on capital and a higher rate of investment. So how do we reduce the private sector's borrowing costs for sustainable investments?

ERIC: We are going to outline five areas of policy innovation which show precisely how to do this. Then we'll return to governance, oversight and regulation. If we do cause funding costs in electricity generation to collapse, we need to ensure the investment happens and returns don't just get siphoned off by vested interests. We need to outline best practice and how to harness technology to make the process open and transparent. Finally, we need to wrap up this discussion with some reflections on how enduring this interest rate environment is likely to be, and how we might think about raising taxes if interest rates were to rise significantly.

CORINNE: We are in the midst of a crisis. It's one that doesn't have the clear and present danger of collapsing stock markets or thousands of people dying from the spread of a disease, but the psychological mindset needs to be equivalent. We have five proposals for collapsing funding costs for sustainable electricity investment: targeted lending, bank regulations, asset purchases, loan guarantees, and a green national endowment. Each one of these proposals treats climate change like the emergency it is, and aims to create EPICs with smart regulations. Let's start with central banks. Monetary policy is an area you have written about and researched for many years. It is one of the most powerful parts of our economic system. How does it work?

CHEAT SHEET 2

When interest rates are negative you get paid to borrow. **Q:** *What?* **A:** Yep, you borrow, and instead of paying interest, the bank pays you! **Q:** *Where can I get hold of these negative interest rates?* **A:** You can't, but in most of Europe, and parts of Asia, your government can. **Q:** *So why don't they borrow like mad?* **A:** They can go one better than that – borrow like mad, invest in climate infrastructure and make money.

ERIC: Most people understand that central banks set interest rates, and many will have heard about things like quantitative easing (QE).[8] They may also know that during the financial crisis of 2008, central banks provided trillions of dollars of support to the global economy and the commercial banking system, and again during Covid-19.[9] Central banks create the electronic money at the heart of our economy, and their job is to create enough money to maintain a low and stable level of inflation over the medium-term.[10] In most countries the central bank is also responsible for regulating the private commercial banking system. It determines the rules under which all major banking institutions operate.

CORINNE: Everything we propose is consistent with the objectives of keeping inflation stable and the financial system secure.[11] Our argument is that these policies will help central banks fulfil these roles, particularly in areas where they have repeatedly struggled.

ERIC: There are three levers the central bank controls, which can be used to create EPICs to accelerate investment in clean energy. The first are called "targeted lending programmes", the second is the regulatory framework for the commercial banking system, and the third is through asset purchase programmes. Let's take them one by one, starting with targeted lending programmes.

CORINNE: In order to illustrate how interest rates and targeted lending can be harnessed as EPICs, let's consider the example of a wind power generation business considering a new investment project. Currently businesses in this sector borrow either from banks or by issuing bonds, effectively borrowing from financial markets. In most of the world, the cost of borrowing is likely to be somewhere between 1 and 5 per cent, depending on the geography and the perceived risks of the borrower. This will limit the investment opportunities that firms consider viable. To make financial sense, any investment has to pay back these interest costs, on top of all other costs, before it generates any return. So if we can bring these interest rates down by several percentage points or more, in some cases even into negative territory, the number of investment opportunities would greatly expand and we could get a huge acceleration in investment in wind power generation.

ERIC: That's exactly right. As you've said, in parts of the world, such as most of Europe, Japan, the UK and the United States, official interest rates are already close to zero or even negative. What we are suggesting is that central banks lend to commercial banks at steeply *negative* interest rates on condition that those funds are lent on to fund new investment in sustainable energy infrastructure, and also contingent on that reduction in the cost of funding being passed on to the end borrower. This is an EPIC targeted lending programme. Are central banks able to do this? Would it work? Few experts doubt that energy infrastructure spending is sensitive to the cost of capital. It is not the sole consideration, but it is a powerful factor. What may surprise people is that a number of major central banks are already making targeted loans at negative interest rates, and some are already directing these lending facilities to climate infrastructure.[12] Most economists I know are unaware of this, not to mention the public and many environmental campaigners.[13]

CORINNE: Let's turn to some of the objections. I want to start by returning to the issue of inflation. For most of the past 20 years central banks have been trying to raise the rate of inflation, which is why they have already embarked on these programmes, but there are some arguing that this long-term trend has been reversed by the pandemic.

ERIC: I am sceptical of this perspective. I think most of the causes of current inflation are due to temporary, Covid-related, bottlenecks which will resolve over several years, if not quicker. Of course, I may be wrong, in which case further monetary stimulus will not be required and our other policies will need to do the heavy-lifting on climate. Circumstances may also vary by geography. For example, Japan is still in deflation.[14] The Bank of Japan has been adamant that it wants to end deflation and create more demand in the economy. It should consider pricing its existing energy investment lending programme at −2 per cent or −3 per cent. That would change the timeline or rate of investment in sustainable energy in Japan.

CORINNE: So this is a policy which depends on the outlook for inflation. If we start to see sustained high levels of inflation in our economies, we may not be able to deploy it. It may also mean it's more pertinent to some countries, and regions such as Europe, than others. Although if another global slowdown occurs, which is a high probability on a five to 10-year view, it should be rolled out everywhere.[15] What about more fundamental objections, that these are subsidies, and that this is not the role of monetary policy?

ERIC: I often hear this objection, but it is not really valid. Central banks have embarked on these initiatives in order to fulfil their mandates. Targeted lending programmes were actually introduced to create more demand and a more stable inflation environment, and they have helped. Also, it is reasonably well-recognized across the world that central banks should support national policy objectives as long as it does not hinder their primary mandates of keeping inflation stable. Given a climate emergency, there is a first-order justification for prioritizing sustainable energy investment.

CORINNE: Is it ok for monetary policy to favour certain sectors of the economy?

ERIC: All forms of monetary policy favour or punish some sectors or segments of the population. For example, very low interest rates in much of the developed world have benefitted the property sector and contributed to high property prices and stock prices.[16]

CORINNE: Which begs the question, why not be transparent and act consistently with the other goals of national policy?

ERIC: Agreed. Monetary policy also affects the distribution of income and wealth. House prices, stock prices and incomes are also affected by interest rates. Under our proposal, these effects would be transparent and there would be proper oversight.[17] That is progress. Doing this in a manner consistent with sustainable growth is compelling.[18]

CORINNE: Ok, so we've described how targeted lending facilities work, how they can be repurposed and supercharged. Let's look at the two other approaches to supercharging monetary policy: commercial bank regulations, and asset purchase schemes.

CHEAT SHEET 3

Banks lend to households, small businesses, big businesses, green businesses and . . . polluting businesses. Q: *What if they stop lending to polluting businesses?* A: Those industries would shrink in size. Q: *So why don't they stop lending to polluters?* A: Exactly.

ERIC: The reputation of banks was damaged by the financial crisis in 2008. As the lifeblood of the global economy they have a huge opportunity to play a positive role in the climate crisis. The changes occurring in financial markets, which we have discussed earlier, are already pushing the largest banks in the world in the right direction. Shareholders are looking increasingly closely at "financed emissions", greenhouse gas emissions financed by loans and investments. This provides a significant motivator, and is another reason why the policies we are proposing here can be very powerful. Banks want to finance sustainable investments at scale. National central banks need to encourage this process with regulatory powers.

CORINNE: Let's talk very briefly about capital requirements. Most people don't know what they are, but they affect interest rates, so they matter.

ERIC: These are the rules set by the central bank which determine the parameters within which banks are operating.[19] Usually, riskier lending has higher capital ratios and therefore higher interest rates.

CORINNE: So another way that central banks could help supercharge sustainable lending is by altering capital ratios. For example, we want the capital requirements to be far lower on green mortgages, such as those that require households to make an investment in energy efficiency in return for lower interest rates, than on conventional mortgages. And we want to raise the capital ratios on emissions-intensive industries, to restrict their access to credit. This lever could be used by central banks, across multiple sectors.[20]

What about the final lever which central banks can use: asset purchases, or quantitative easing (QE), as it's called.[21] We don't need to get lost in the weeds, but the bottom line is that central banks, which are public institutions, now own trillions of dollars of bonds and equities issued by private companies.[22] In effect, these are assets which we, the general public, own, and central banks are custodians acting on our behalf. Can these huge portfolios be used to promote adherence to international best practice on climate finance?

ERIC: Absolutely. It seems very odd that we expect fund managers in the private sector to be investing with ESG standards, as we discussed earlier, but central banks who represent us are not acting as good citizens. Central banks should in fact be at the forefront of better governance, and setting a higher bar than private sector investors.

CORINNE: Many climate campaigners and members of the public may be unaware of the existence of these tools, let alone their potency as policy levers. A final point I want to make on monetary policy is how quickly central banks can act. Central banks can simply have a conference call and make a change overnight. How quickly do you think the measures you are describing could be implemented?

ERIC: This is a really important point because a significant part of our proposals is to dramatically shift timelines. What we've described here could be done in a matter of months, and it could accelerate the rate of investment spending and the prospects for energy sustainability across the developed world. It is borderline negligent that these policies are not already in place. What we are outlining would provide both huge economic stimulus to weak economies, but also transform the sustainability of our energy sectors and electricity provision.

CHEAT SHEET 4

Creating a sustainable economy involves building wind farms, solar farms, energy storage infrastructure, charging stations for electric cars, and lots of new innovative businesses we haven't even thought of. **Q:** *Who's going to own them?* **A:** If we provide the capital, we can take a stake in these businesses. **Q:** *Wait, who is "we"?* **A:** The government. **Q:** *We own the government, don't we?* **A:** Yep.

CORINNE: We shall discuss later how our policy ideas can generate moral redress for the causes of climate change, but even more importantly, create jobs and prosperity.[23] The UK, where we live, has some of the worst income and wealth inequality in Europe, which is also reflected in many other measures of well-being, such as poor health outcomes. The US, the world's second largest emitter, has a similarly extreme picture of wealth, income and social inequality. One idea we've discussed at length, which you and Mark Blyth proposed in *Angrynomics*, is that of a national endowment. It's an idea that gets me excited because we can tackle the energy transition at the same time as wealth inequality. Let's dig into it.

ERIC: The idea of a green national endowment draws on existing models of sovereign wealth funds, but with a twist. The implications of low levels of interest rates for sustainable investment is a central theme of our discussion. The novelty Mark and I introduced in *Angrynomics* is the idea that the historically low interest rates at which governments can borrow provide a huge opportunity for the state to create wealth. So far we have focused on how central banks can reduce interest rates further for the private sector, through targeting lending and other tools of monetary policy at sustainable electricity generation, using the commercial banking system which is already incentivized by markets to transform its asset mix in favour of sustainability. The innovation we are suggesting now, is that the state can use its balance sheet to create wealth in green assets, and we can also distribute this wealth to those in the population with no assets. Broadening asset ownership without raising taxes can also serve as a very powerful political message, and could also transform hearts-and-minds with respect to climate change.

CORINNE: Rather than fearing for our children and grandchildren's future, why don't we create green assets to provide them with an inheritance? For readers who haven't read *Angrynomics*, and aren't familiar with the idea of how the government's low cost of debt can create wealth, and how this could be distributed to those in the population who have no assets, can you explain how this works.

ERIC: Let's use the analogy of buying a property and then renting it. Imagine you could take out a mortgage with an interest rate which is fixed for 20 years at 0 per cent. As long as the property you purchase generates a positive rental income, after all costs, you will create wealth through time. If the property pays a net rental income of 5 per cent in 20 years, you will have repaid the underlying loan over those 20 years, and are left with 100 per cent ownership of an asset. Now the private sector can't borrow at fixed rates of interest for ten years or more at zero rates, but the state can – this is essentially the position for many nations across the developed world currently. What we argued in *Angrynomics* is that the state should issue debt right now to fund a national endowment, which would be a new independent, national institution mandated to generate something like a 4 per cent real return over the medium term.[24] There are lots of global models we can learn best practice from, such as the Norwegian sovereign wealth fund, which we have talked about before and which has generated a 6 per cent compound return over 20 years, but there are also not-for-profit endowments, such as Harvard's, or the Wellcome Trust in the UK. If we consider the Norwegian fund, two thirds of its current value has been generated through returns.[25]

CORINNE: Let's just pause on that, because it is a critical statistic. It demonstrates the power of money to make more money.

ERIC: Exactly. If the Fund had funded its initial endowment with debt fixed at zero interest rates, it would now be able to repay that debt and would still hold assets equivalent to more than 100 per cent of Norway's GDP. That is the power of compound interest, and the opportunity afforded by close to zero interest rates.

CORINNE: So, we are clear on the economics of how an endowment fund can create value. But how do we ensure it contributes to the green transition?

ERIC: So far, most of our policy prescriptions have focused on altering the interest rate on the debt side of the private sector's cost of capital. The investments that the national endowment makes will be more equity-like. We should expect these to generate returns lower than those required by the private sector, but higher than the state's cost of funding. That suffices to create wealth, and makes economic sense at multiple levels. If we create a fund by issuing government bonds, we can then use the proceeds to create a national endowment which invests in green assets – including equity-like investments in charging infrastructure, storage, venture capital funds looking for technological solutions to the challenges related to energy efficiency in buildings, or cement manufacture. At inception, the green endowment fund would invest a majority of its capital in a diversified portfolio of publicly-listed assets, replicating the models of existing endowments and sovereign wealth funds. It could act as a best practice global ESG-fund, with a portfolio return objective of, say, 4 per cent in real terms. Through time, over the next 10, 15 or 20 years, as the fund's value compounds, we can then distribute ownership of these assets to those parts of the population with less wealth or asset ownership.

CORINNE: The effects of low interest rates is not unlike the discovery of North Sea oil.[26] Before the 1970s, Norway was a relatively low-income, agricultural based economy. It was the discovery of oil which seeded the Norwegian sovereign wealth fund. The current interest regime is like that moment in time, a form of luck that enables us to create a state-owned stock of capital. We should clarify a number of issues. People get very preoccupied by the idea of more government debt, which has risen further during the pandemic. People worry about investments in the stock market, which come with risk. And there are lots of concerns about governance. Do we *trust* our governments to do this?

ERIC: These are all legitimate concerns, which need to be addressed carefully. There is a high degree of confusion around government borrowing. We are advocating a huge expansion in the state's balance sheet to tackle climate change. This also occurred during the global financial crisis and during the pandemic, but there is a critical difference. We are also advocating the creation of assets which the state will own. We are not advocating nationalizing anything. We are saying the state needs to act as a financial actor because its cost of borrowing is so low. The private sector will do the building and management of the assets we create, but the state needs to provide guarantees, borrow money, and take equity-like investments across many sectors in order to create wealth for the public and accelerate the green transition.

CORINNE: We shall come back to this issue when we discuss government guarantees and fiscal policy, but the idea of a national endowment illustrates that there is a critical difference between borrowing to buy assets and borrowing for current consumption. Borrowing to buy assets creates value to the state through time, if the interest rate on its borrowing is lower than the return on its investment. It's as simple as that.

By contrast, borrowing and spending the money on consumption leaves the state with a liability. This feature of fiscal policy needs to be at the centre of the debate. Everything we advocate involves borrowing at an interest rate lower than the return on the investment it funds, thereby creating wealth for the state. But, what about borrowing and investing in public securities including equities? Isn't this risky? What if there's another stock market crash?

CHEAT SHEET 5

Q: *This all sounds too good. Our interest payments are the lowest for decades, we can invest at a higher return and make money, and we can take a share in the wealth we create. There has to be a catch?* **A:** Some people will get nailed. **Q:** *Who?* **A:** People who own businesses that produce coal, oil and natural gas. **Q:** *I don't know any of them.* **A:** Nor does most of the planet. In fact, they're tiny in number, and mostly very rich.

ERIC: This is why we need to take a 20-year time horizon for wealth creation. Not in terms of how quickly we can super-charge the economy, that can be immediate. But in order to be reasonably confident that we can create wealth for the nation, a long-term perspective is essential. All of the major funds we have referenced have had periods of negative returns. But over decades their returns have compounded at quite high rates of return.[27] Stock markets do crash, but they also recover. The bigger concern, I believe, centres on governance. With all policy interventions along the lines we are advocating, the biggest risk is that vested interests both within the political system or the private sector capture programmes.[28] Fortunately, we have lots of examples now of how to tackle this risk and what best practice looks like, but this needs to be a more integral part of any debate on public policy.[29]

CORINNE: Before we move on to the broader issue of how *fiscal* policy can supercharge the nation, I want us to briefly counter the accusation that our perspective is Panglossian. We have a unique opportunity to create wealth by investing in the green transition, and this can be done in a way that enables asset ownership to extend to those in our societies with few or none. At the same time, however, there will be losers. Someone owns the assets which will be stranded.

ERIC: The assets that will see their values destroyed, effectively the polluting assets such as coal-fired power plants, are owned by a tiny percentage of the population. Less than 0.1 per cent of the global population owns coal, oil, and natural gas assets. Gas is a critical transition fuel, but it will eventually lose its value, if on a slower timeline than coal and oil. These assets tend to be owned by theocracies, thugocracies, and a fraction of the "1 per cent". The assets we want to create with our national endowment will actually replace these and render them redundant, particularly when seen in the context of the other policies we are advocating to accelerate capital flows to green investments. Importantly, the national endowment would be distributed to those who don't have assets.

CORINNE: This point is critical. We hear so many fears about the costs of transitioning to a sustainable economy and stopping climate change. There *is* a large cost for the decarbonization of our economy. Currently the estimated market value of what will become stranded assets, oil reserves, coal reserves, publicly listed carbon consuming businesses is equivalent to approximately $4 trillion.[30] This is concentrated in the hands of very few, and disproportionately focused in some regions, such as the Russian and Saudi Arabian economies. If these countries want sound medium-term strategic policies, they should be rapidly decarbonizing and diversifying into green technologies.

ERIC: To be clear, there are concrete humanitarian challenges facing some communities dependent on the fossil fuel economy. A robust national climate strategy has to have a plan to reskill and invest in these communities. Overall, however, a relatively small share of global employment is tied to the fossil fuel industry and there already is increasing demand for workers in sustainable industries, which will intensify.[31]

CORINNE: We are describing a large increase in state borrowing. We have addressed some of the key issues, such as the critical distinction between borrowing to invest, which creates wealth, and borrowing to fund current spending, which leaves us with more debt. We have also described how monetary policy can create EPICs, and how the state's balance sheet can be harnessed to create green assets. But not all government-funded EPICs are forms of capital expenditure. We have also argued that we should use tax exemptions, or even negative taxation to sharply alter the relative price of green substitutes in every major sector of our economy, from electric vehicles, to green steel, to plant-based burgers. We have argued that EPICs will change behaviour, and that a positive incentive relative to the status quo is more politically viable, more effective, and will win popular support. But all of this results in lower revenue to the government, or does it? That's the first area I want to discuss and what the implications of this might be. The second question is a more specific one about scaling renewable electricity infrastructure. So far we have discussed electifying transport and industry. How might nations apply EPICs to buildings, which often don't get enough attention.

CHEAT SHEET 6

Q: *Ah, but if every country does what you suggest, and there's an economic boom, then interest rates will have to rise, and what will we do then?* **A:** The first thing to remember is that our governments borrow at fixed rates of interest for up to 30 years. But you are still right, there could be a boom and inflation – no one can say with certainty. **Q:** *What then?* **A:** Well first, that would be good. Emissions would be falling, lots of jobs being created and incomes rising. But we might have to raise taxes. **Q:** *OK, so who will pay then?* **A:** That will be up to all of us, and who we elect.

ERIC: Ok, let's start with the money, and then talk about buildings. Our ideas suggest it's a strategic error to focus on raising taxes. If the population associates climate change with higher taxes, when most people already feel that they are struggling to make ends meet, there is an immediate barrier to progress.

CORINNE: We want the population to associate the green transition – rightly – with investment, wealth creation, job creation and regional development. EPICs, by definition, are positive incentives. In every sector, we want the green option to be significantly lower cost. This must be done through tax exemptions and negative taxes.[32] At the same time, we clearly need a plan for intelligent management of the public finances.

ERIC: The first step is to be very humble about our ability to predict outcomes. If the green transition is implemented along the lines we are describing, it is likely there will be strong economic growth. That is usually what happens if there is high investment spending, lots of innovation, and the rapid emergence of new industries and sectors.[33] It is possible there will be a self-sustaining green boom. Alternatively, our policies may be so successful that governments need to raise taxes, because deficits are too large and inflation and interest rates are rising. I suspect this is unlikely, but it is possible. In these circumstances, fiscal policy may need to respond and the prescription is straightforward: raise taxes in an efficient manner. We should absolutely not be raising "green taxes". We should use the tax system to address problems which are entirely independent of the green transition, such as income and wealth inequality, corporate tax avoidance, and uneconomic rents. It is an extremely bad idea, economically and politically, to attempt to match and balance the "cost" of creating a sustainable economy. It is also impractical. We do not know how the major economic forces will evolve over the next decade or so, and aggregate fiscal policy needs to be responsive. If we do need to raise taxes, we should approach this problem independently and in order to meet other social goals.[34] Climate change is an independent priority.

CHEAT SHEET 7

Q: *I don't have the £6k needed to pay for a heat pump?* **A:** Tell me about it. But what if the bank pays for it, adds the value to your mortgage and cuts the interest rate, so you pay less each month?

CORINNE: So, what can we do to tackle emissions from buildings?

ERIC: Making one's home, or commercial real estate, less emissions intensive shouldn't face much resistance, because incentives are aligned.[35] For example, if we invest to improve energy efficiency we can pay off the upfront cost over time by saving on our energy bills. In practice, however, this has proved very difficult, far more so than the relatively rapid progress some nations have made on greening electricity. A number of countries have introduced subsidies to reduce or eliminate the upfront cost, but uptake is almost universally disappointing.[36] There isn't much evidence globally of effective policies to shift energy efficiency of residential or commercial buildings.

CORINNE: Why do you think this is?

ERIC: First of all, there has been a failure to adhere to the golden rule of policy-making: keep it simple, make sure it's effective and make it non-partisan. The second reason is a failure to use EPICs. Where incentives have been extreme, for example in domestic solar panel installation, take up has been rapid.[37] The problem has usually been government inconsistency. It is essential to acknowledge that the practical problems are very substantial. Policies involving high upfront cost and considerable inconvenience to home-owners, are not going to work, despite potential long-term savings.

CORINNE: Significant improvements in energy efficiency within residential properties often require a large capital outlay, often as much as £20,000 per property.[38] Why would households take on large capital expenditure like this? On top of the upfront cost, you have to minimize inconvenience. We know this from behavioural psychology.

ERIC: The evidence from EPICs suggests we have to go beyond mitigating the costs of upfront capital outlay or inconvenience. We have to proactively create large incentives for people to be motivated to make these investments. These incentives need to be financial to shift initial behaviour. Eventually social stigma will play a role and you'll be a pariah if you are the only house in the neighbourhood that isn't heated by solar panels or a heat pump.

CORINNE: Before we get to the role of EPICs, we need to be clear on our objectives. This is an area where insufficient clarity of priorities has been a problem. For example, across most of the developed world, buildings directly and indirectly – mainly via electricity consumption – are responsible for almost one-third of CO_2 emissions. The indirect source of emissions, electricity, accounts for around 70 per cent of total buildings' emissions.[39] This is precisely why we have prioritized making electricity generation sustainable. If electricity is emission-free, the carbon footprint of many buildings would collapse. This poses an obvious tactical question: if we are going to invest in buildings to reduce emissions, should we target energy efficiency, or should we repurpose existing infrastructure which runs on fossil fuels, to be powered by electricity?[40] As a practical matter, there is little doubt that shifting to sustainable electricity is likely to be the most effective policy to collapse building-related emissions, but these objectives do not have to be mutually exclusive, and the specific conditions vary significantly by country and within regions of the same countries. For that simple reason, a multi-targeted approach is wise, with significant variation by region, and types of building. But once priorities are designed taking these local factors into account, if we want change fast, we have to deploy EPICs. We've developed some ideas about how these could be deployed for building emissions reductions, do you want to describe them?

CHEAT SHEET 8

Q: *I live in rented accommodation. How can we get the landlord to invest in insulation or electric heating?* **A:** Banks can offer landlords better mortgage rates for greening properties, too. We will also need to regulate them.

ERIC: We have had conversations with commercial banks, regulators and policy-makers on this topic. Our conclusion is that a combination of green mortgages, smart regulations and negative taxes may be the most powerful levers. As with all of the policies we are describing, a great advantage of this being a global challenge is that many countries are already experimenting, and learning what works.[41] It is imperative that all policy designers look at what has been done elsewhere, and copy from the examples of success. We have to accept that in many areas this will be an iterative process of failure, learning and then success, but there really should be more learning from past failures.

CORINNE: Let's expand on this. Firstly, what is a green mortgage?

ERIC: Green mortgages have preferential rates of interest linked to emissions reductions.

CORINNE: Who has been most successful in developing effective green mortgage products?

ERIC: So far, I think the large US government-sponsored mortgage lenders have probably been the most successful. They tick at least two of our boxes. They use EPICs, so the interest rate on a green mortgage is lower than the alternative, and they are exploiting the preferential pricing in financial markets, becoming the largest single issuer of green bonds in the United States.[42] There is lots to learn from this model.[43]

CORINNE: We also want to harness property taxes as EPICs. Most countries have multiple forms of property tax, either on transactions, or as recurrent forms of wealth tax.[44] This offers lots of options. There should certainly be two tiers of property tax. Governments could offer five or ten-year exemptions to those who make capital investments in emission reductions. Similarly, property purchases which are financed by green mortgages could be made exempt of transaction taxes. Again, this increases the probability of take-up and changes perceptions of what the green transition means for citizens.[45]

ERIC: EPICs are important, particularly when we are trying to mobilize changes in behaviour across the household sector, but where there is social or political viability, smart regulations are also essential.

CORINNE: We explored this very early on in our discussion. The most effective way to deal with many forms of pollution in the past was to simply ban them. Regulations played a critical role in the fuel efficiency of vehicles, for example.[46] In new buildings and the rental market, regulations play a key role. This is again an area where experimentation across the world will deliver many lessons. The European Union has embarked on a very ambitious set of regulations for commercial properties, as has the UK.[47]

ERIC: Before we wrap up this part of our discussion, I want to return to a key issue we raised, which is essential to execution and public confidence in what we are advocating, particularly in the design of electricity infrastructure projects – governance. What are the dimensions for good governance that we need to take into account in designing these policies?

> ### CHEAT SHEET 9
>
> **Q:** *If the government is paying for lots of this investment, how do we stop politicians from feathering their own nests? Or the nests of those who put them in power?* **A:** This is a real challenge, depending on which region or country we look at. We do need some level of trust, but we also need robust transparency and public scrutiny. We all need to police this as individual activists.

CORINNE: When we look at the issue of optimal governance, or how to minimize forms of capture or corruption in areas of public policy, there are three broad axes which are relevant. The first is transparency, which involves making public all the data in the procurement process between government and the private sector. So radical transparency is a powerful force to improve governance. Technology is a major enabler here. It means private sector bidders can see why they have lost a contract and what terms were awarded to the winner.

ERIC: NGOs and other civil society representatives can scrutinize the distribution of public funds, and public officials can use the data to improve project efficiency. The second axis is institutional design and agency. Which institution is responsible for disbursing funds? This will vary hugely by country, and requires a pragmatic assessment of institutions' tendency to capture. For example, in a nation or a region where local government is subject to capture by the local corporate sector, it may make sense to provide incentives directly to households.

The final consideration is to combine local accountability with distant, or arms-length, disbursement and oversight. The irony is, corruption requires a high level of trust. Often it is local connections that result in capture. At the same time, there is critical local knowledge. The European Union is an important example. EU projects may be less susceptible to local capture if they are overseen centrally. But local communities may have an important role in ensuring efficiency.

CORINNE: Bill Gates apologises for talking about policy. We have done little else. But towards the end of his book he presents some evidence detailing how government incentives in Japan, Germany, the US, China and Denmark transformed the solar and wind industries.[48]

ERIC: These are not arbitrary success stories. Gates doesn't realize it, but these are all EPICs. That is why we have described how an EPIC nation would work. Most people don't know much about monetary policy, fiscal policy, and bank regulations. The bottom line is that they can all be repurposed to act as EPICs to finance a green investment boom. Businesses rely on banks to provide credit, particularly for long-term capital investments. There is an immediate link with our number one national objective, electrification and sustainable energy generation. Achieving this goal requires huge investments, across renewables, transmission and storage infrastructure. So how can commercial banks and the central bank accelerate the process using EPICs? Central banks can offer low, or negative interest rates for the commercial banks to pass on to sustainable infrastructure projects, expanding the number of projects that get funded. Governments can provide guarantees, make direct loans, take equity stakes and provide funding for research and development.

CORINNE: The conventional wisdom that society has to pay a big price for the green transition is a misconception. We have the lowest interest rates in history and a wealth-creating investment boom in the making. If we look at economic history, phases of huge capital investment have tended to transform living standards. Although we are destroying part of our economy, such as legacy energy systems, we are also creating highly valuable assets in sustainable electricity generation, in electricity transmission, in green buildings, in electric vehicle manufacturing, in battery technology and in charging infrastructure. Jobs will be created and there will be spillover effects from the R&D and investment spending.[49] By supercharging the greening of the globe, we are also supercharging an improvement in living standards and the quality of life.

6

Supercharge the world

CORINNE: We have tried to ground *supercharging* in pragmatism. We have proposed policies consistent with an honest take on political, social, psychological and economic reality. Before we describe our proposals to supercharge the world, we need a blunt assessment of the constraints on global policy-making. Global politics often feels like a chaotic set of events that don't make sense, but we can observe some very common threads. Our friend, Mark Blyth, describes the world pithily: "The era of neo-liberalism is over. The era of neo-nationalism has just begun."[1] You and Mark discuss the emergence of the era of nationalism extensively in *Angrynomics*. One way to make sense of politics is to think in terms of trends and the incentive structures that politicians are operating under. The simplest incentive politicians have is to get elected. Let's discuss what neo-nationalism is, and how it influences that incentive to get elected.

ERIC: In representative democracies where elections are often swayed by the votes of relatively small minorities, using nationalism to swing a decisive minority of the electorate is an obvious strategy for politicians to exploit. It is also striking that a political system like China's, which does not hold national-level elections, also harnesses anti-other ideology (usually anti-America) to unify and motivate the domestic population. Tribalism is a motivating instinct which is very easy to switch on. We see this within advanced economies like the UK, we see this in former communist countries like Poland and Hungary, and we see this in emerging economies like Turkey and Russia. Trump's "Make America Great Again" adds nostalgia to Berlusconi's slogan, *"Forza Italia"*. Tribalism is an electoral and political strategy either to disguise the true agenda, or to compensate for an intellectual vacuum.

CORINNE: How does this compare to the era of neoliberalism?

ERIC: The era of neoliberalism broadly refers to a period from the early 1980s until the mid-2000s. It was defined by a shared set of economic beliefs across the western developed world, and from the late 1980s, across former communist countries as well. This period was defined by a consensus in economic policy, devolving control of industry to the private sector, increasing the role of markets in labour and product markets, and retaining a narrowly-defined role for the state.

CORINNE: The role of the state being technocratic governance of economic policy, with varying degrees of welfare provision.

ERIC: Exactly. The period is also defined by a process of greater global cooperation favouring the free movement of people, capital, and goods and services, encouraged through international trade agreements. The expansion of the European Union and the World Trade Organization are defining institutional features, which exemplify a voluntary dilution of national power in favour of international cooperation.

CORINNE: We still have a relatively free flow of capital, goods and services, and substantial global migration of people.[2] There was a lot of noise about trade wars with Trump, and at the periphery with Brexit, but not much that is really concrete. So what has *actually* changed?

ERIC: Political narratives have probably changed more than institutional reality. The most concrete shift since the end of the Cold War is probably the decline of the idea of the market economy being an end-in-itself, and advocacy of free markets sufficing to motivate critical parts of the electorate. Trump is very, very different to Reagan in this regard. At the same time, many historians and political scientists will reasonably contend that national self-interest has always been the dominating force in international politics. Even during the era of neoliberalism, adherence to true international governance was substantially a charade.

CORINNE: For our purposes, we are taking as given that national decision-making on global issues will usually be premised on self-interest, or at best enlightened self-interest. When we reflect on countries critical to the energy transition, such as the United States, China, India and Russia, this is an important perspective.

ERIC: Yes, climate change will not be addressed by ethical motivation alone. Our policies have to be based on mutual reward. Rogue states should also be recognized as such, and we need tactics for dealing with them. Russia is a good example, which we shall return to.

CORINNE: This backdrop echoes the perspective of both Dieter Helm and Anatol Lieven. Helm's critique of climate progress over the past 30 years references an over-reliance on Fukuyuma's "End of History" paradigm, which suggested we had emerged into a truly universalist, rationalist mode of operating, where nations and individuals were capable of putting the greater good ahead of their own self-interest. Helm suggests that this paradigm misconstrues human nature, has failed to deliver on the climate crisis, and cannot be relied on going forward.[3] The appeal to nationalism is also the central thrust of Lieven's *Climate Change and the Nation State*.[4]

ERIC: I think there are enough appeals to nationalism out there without adding to them. I am not as pessimistic as Lieven, but I take on board the need for realism. There are chinks of light. The common emphasis across the largest economies on climate change reveals that despite the best efforts of nationalists, at some level common purpose and a universal human vision is also irrepressible. Humans are local, parochial and tribal, but also have an extraordinary capacity to be motivated by a sense of the greater good, and an ambition for a better world for future generations. Many of the most forward-looking countries on climate change will have very little direct impact on the outcome, their influence is through example-setting, leadership in technology, ideas and cooperation. Learning from cross-country comparison is an underlying theme in our discussions and a major reason for optimism. There are many technical challenges to electrification, but if almost every country in the world is trying to find answers we can learn from, the probability of success is far higher.

CORINNE: Now we have the backdrop, let's segment the world. First of all, we have the motivated developed world. This includes most of Europe and North America, depending on who is in power, and large parts of Asia. These are high-income countries, whose governments are broadly committed to tackling both national and international emissions and are reducing the carbon intensity of their economy, if not reducing absolute emissions. Then we have the emerging economy equivalent, governments which are motivated and committed to emissions reduction, but operating in a different economic context. This segment includes countries such as India, Indonesia, Mexico, large parts of Asia, Africa, and Latin America. China is in a camp of its own, due to both the sheer scale of its emissions and its unique political economy. Finally, we have a trickier category of nations who are not motivated or committed to tackle the climate crisis, and in some cases, the structure of the economy relies on fossil fuels, creating a disincentive to support the energy transition.

ERIC: This final category includes Australia, Russia, Turkey and Saudi Arabia.[5]

CORINNE: We need separate strategies for each of these groups of countries. In our discussion about supercharging the nation, we outlined a strategy for those economies where interest rates are already close to zero and there is no sign of a debilitating financial constraint. This includes the motivated high-income economies of the world and China. Let's explore how to help supercharge lower-income economies, which are motivated to change but face serious financial constraints.

ERIC: Our discussion should provide grounds for optimism, but not false optimism. One of the major concerns people have is that the changes they make in their own lives, or at a national level, are insignificant in the global context. Take Denmark as an example. Denmark has committed to reducing emissions by 70 per cent by 2030, on a 1990 baseline, and has already delivered a 32 per cent reduction.[6] This shows what can be done. Yet Denmark accounts for a mere 0.1 per cent of global emissions. So a Danish citizen or policy-maker motivated to move the needle on global emissions, should be thinking, "What can we do to accelerate the transition in countries like China, India or Brazil?"

CORINNE: How very Danish. Independently of the direct impact on emissions, there are huge economic benefits of electrification for an economy like Denmark. Ambitious countries provide powerful examples to the rest of the world about what can be done and how to do it, not to mention the benefits of technological expertise which can be exported.[7] At the same time, viewed solely through the lens of the direct effects on emissions, Denmark is relatively insignificant. So how might Denmark have an impact on, say, India? India matters far more than any individual country in Europe.[8] Alone it accounts for approximately 7 per cent of global emissions and ranks third after China and the US.[9] Equally important, and in sharp contrast to Denmark, and almost all developed economies, India's emissions are still *growing* rapidly.

ERIC: Explain what is happening in India, so we can work out if a relatively small country like Denmark could help.

CORINNE: India is a very complex emissions challenge. It is one of the most coal-heavy energy sectors in the world, second only to China, and its coal consumption is also growing. There are specific reasons for India's coal dependence. It has vast domestic coal resources, and large steel and cement sectors, which are also coal-intensive. India's politics is neo-nationalist and like many countries, energy policy is significantly motivated by self-sufficiency, ahead of climate objectives. If we take sustainable electricity generation in isolation, India is a great success story. It has delivered extraordinary growth at scale and has ambitious targets.[10] By contrast, if we look at coal consumption and efficiency, India is the world's leading offender.[11] So, there are huge challenges to weaning India off coal. The coal and steel industries are very important in the context of India's political economy and coal resources and heavy industry are concentrated in some of the poorest regions of India.[12]

ERIC: It might surprise people that we don't need to convince India on the merits of transitioning its economy to sustainable energy. Like many countries, air pollution provides a further motivation to transition out of fossil fuels into cleaner renewable energy sources. Approximately 84 per cent of the Indian population is exposed to pollution above the limit of India's national standard.[13]

CORINNE: That's right. What India is already doing with solar is more ambitious than what we see across much of Europe and the United States. Between 2014 and 2020, India's installed capacity of solar power grew 13-fold, to 36 gigawatts.[14] To put it in context, that is around four times Spain's installed capacity, built in six years![15] In fact, when people argue that we are too ambitious on the potential for growth in sustainable energy, India provides evidence of what can be done at speed.

ERIC: Ok, so despite the challenges, Indian policy is focused on reducing emissions intensity. The issue is not how to convince India to move fast, but how to provide support for transitioning its coal industry. How can high-income economies facilitate India's transition?

CORINNE: The strategy echoes what we have argued for elsewhere. EPICs work in India, too. We have argued that part of making EPICs work in electricity generation is in de-risking price volatility, this reduces the cost of capital and results in higher rates of investment. Reducing risk and the cost of debt are two key components of supercharging electricity. There are lessons we should all be learning from regulatory mistakes elsewhere. India negatively impacted investment in wind generation by introducing "market risk" into the tariff structure.[16]

ERIC: So, as in the developed world, regulatory changes to de-risk the electricity sector would help accelerate investment in India?

CORINNE: Yes. But India is also financially constrained. Given the sheer size and carbon intensity of India's energy sector, we need to supercharge the rate at which India transitions from coal to renewables, and this requires capital.[17] So the key challenge for all the developed world, and including Denmark, is how to provide India with capital. At the same time, the Indian steel industry could also be encouraged to change more rapidly through a green sectoral trade agreement, which we discussed earlier.

ERIC: India has an ambitious commitment of reaching 57 per cent renewable share of total energy capacity by 2027.[18] So the motivation is there, with significant caveats. We shall deal later with countries where that will is lacking, such as Russia or Brazil. As you identify, one of the key constraints on the transformation of the Indian energy system is capital, which is linked to financing costs. India will need to spend $1.4 trillion in its energy infrastructure over the next 20 years, 70 per cent more than is currently planned.[19] Most Indian power projects are financed with a cost of capital as high as 10 per cent.[20] Whereas countries such as Denmark can borrow for up to 20 years at less than 0.5 per cent, fixed.[21]

CORINNE: Surely India is far too large for a country like Denmark to single-handedly make a financial impact?

ERIC: You would think so, but you would be wrong. If the Scandinavian countries of Denmark, Sweden and Norway wanted to, they could comfortably finance 10–20 per cent of India's electricity infrastructure needs over the next decade. If we took Europe as a whole, we could finance all of India's power sector investment with less borrowing than we did in one year of the pandemic.[22] And we would be repaid, with interest.

CORINNE: This is the idea behind a "Green Bretton Woods", a central policy recommendation for supercharging the world.[23] A key point is that loan guarantees and financial investments in infrastructure create assets. This is not a *cost* to high-income economies. They will make a return on these investments. In many cases, this will also create additional demand for capital goods exporters in high-income economies, creating jobs and fostering technological innovation.[24] Many European economies are already benefiting hugely from exports of capital goods for the Indian renewables markets.[25] We cannot stress this highly enough. There is a deal to be done: we can significantly reduce the financial cost of transforming India's coal industry, while obtaining a return on our investment and increasing our exports.

ERIC: For readers unfamiliar with the original Bretton Woods, why is it a helpful reference point, beyond the reference to trees?

CORINNE: Bretton Woods was a system of monetary stabilization set up after the Second World War to address one of its perceived causes, global economic instability. Keynes, one of the intellectual leaders behind Bretton Woods, realized that we needed international cooperation to solve a global problem. The thesis that international capital flows can be highly destabilizing is uncontroversial. As is the idea that economic instability can lead to political extremism and consequent human suffering.[26] There are clear parallels with the climate crisis.[27]

ERIC: In July 1944, all 44 Allied nations signed the Bretton Woods agreement, committing to the stabilization of exchange rates based around gold reserves and the dollar. Measured by the absence of a major global financial crisis during this period, the Bretton Woods agreement can be deemed a success. It reveals what global cooperation can achieve. Now we have a different challenge. Humans usually pay for insurance *after* disaster strikes. Bretton Woods was a response to a catastrophe. We are attempting something more ambitious, international cooperation over capital flows in order to *prevent* a catastrophe, and against the backdrop of rising nationalism and "mini-catastrophes" such as wildfires and extreme weather events.

CORINNE: So how would a Green Bretton Woods work in practice?

ERIC: The objective is to harness the financial resources of the high-income economies to accelerate sustainable electricity generation in motivated low-income economies, such as India which will serve to phase out coal more quickly. Two institutional arrangements need to be supercharged. The first, appropriately, are residual institutions from the original Bretton Woods agreement, namely the International Monetary Fund (IMF) and the World Bank. The second avenue is via existing or new bilateral institutions, including export credit agencies (ECAs), which need supercharging, and emerging green investment banks or green national endowments, which should be rapidly growing their international portfolios.

CORINNE: How realistic is repurposing the IMF or World Bank for the next ten years to be 80–90 per cent focused on the climate transition?

ERIC: These institutions are often hindered by politics and bureaucracy. We need an agile, unencumbered institution to deliver at the scale and pace required. I am not sure that the World Bank is up to the task, but there could be a hack within the IMF. To very little fanfare, and unbeknown to most, the IMF has proved a useful vehicle for unleashing vast financial resources in previous moments of crisis, such as the global financial crisis and Covid-19. It has a little-known magic power, which was designed in 1944, and has survived due to the brilliance of its design.

CORINNE: Go on.

ERIC: We framed this discussion around an acknowledgement of global neo-nationalism. A deep irony is that an almost 80-year old global institution continues to run counter to this. There is a silent, unremarked upon institution which has in fact pooled the sovereignty of the United States, China, the euro-zone, Japan and the UK. It is a form of global money, and the IMF is the custodian.[28] Now bear with me. We're going to get technical. Under the Bretton Woods regime, a financial device called a "special drawing right" (SDR) was created.[29] SDRs are a global currency which can only be traded among central banks, and a limited group of eligible institutions. They amount to creating money out of thin air, but are also an act of generosity. Here's how they work. The board of the IMF agrees to issue a certain number of SDRs, and each SDR entitles the holders to receive an amount of US dollars from the US central bank, some euros from the European Central Bank, some renminbi from China, some yen from Japan and British pounds from the Bank of England.[30] If, for example, the Central Bank of India receives an SDR from the IMF, this in effect means that it can convert this into 42 cents, 0.31 of a euro, etc. In other words, it has access to money which is created by the seven central banks who fund SDRs. This is an act of generosity because the seven countries could have created that money and used it for themselves. Now SDRs aren't given away, but they can be drawn upon by virtually any country in the world, under certain conditions and with approval from the IMF. In 2021, SDRs have been used, both bilaterally and under IMF programmes to help many poor countries to cope with the financial fallout from the pandemic.[31]

CORINNE: The cumulative issuance of SDRs leaves the IMF with total resources of around $1 trillion and, in principle, most of this could be deployed. Could this capital form the basis of a global climate fund?

ERIC: Absolutely. As we made clear in our discussion of supercharging the nation, the high-income economies do not face economic obstacles to transform their electricity sectors, it is a failure of will. But many lower-income economies do face serious financial constraints. South Africa is a prime example. It accounts for around one third of Africa's emissions and slightly more than 1 per cent of the world's. It is also heavily reliant on coal, which fuels around 80 per cent of South Africa's energy needs. South Africa is highly amenable to solar and wind power, the challenge, again, is a shortage of capital. South Africa's electricity sector has had enormous difficulties over the last decade that are primarily due to under-investment. A smart global green fund, with something like $500 billion to $1 trillion at its disposal, administered by the IMF, could target high-impact, financially-constrained economies, like South Africa, and facilitate an ambitious repurposing of its coal-fired electricity generation with solar. South Africa wants to do this, but it lacks financial resources.

CORINNE: Currently, the IMF and the World Bank are not deploying at sufficient scale to genuinely impact climate change.[32] What needs to happen politically to release the financial resources of existing SDRs to finance a focused drive towards sustainable power generation in low-income economies?

ERIC: The Biden administration could almost certainly do this. The smart political tactic would not be to request a new round of SDR issuance, but rather a repurposing of the Covid resources, once the recovery is complete. This would probably amount to something like $500 billion, which if targeted smartly could make a huge difference. Again, it is critical to point out that these are financial resources which would be used to invest in creating power-generating assets. If done properly, these assets generate positive returns. No one is giving money away, we are pooling our collective financial power to create assets and change the trajectory of climate change. What could be truer to the ideals of the original Bretton Woods agreement?

CORINNE: So the first phase of a Green Bretton Woods would see the financial resources of SDRs unleashed to provide loans or loan guarantees for sustainable electricity generation in financially-constrained low-income economies. This is probably dependent on US leadership, and requires that the IMF ups its game. What about the bilateral options?

ERIC: There are a number of bilateral options. The strategy should be to empower and supercharge existing institutions, and turn the focus to the climate challenge. For example, most developed countries have institutions known as export credit agencies.[33] Export credit agencies have been designed to deal with the precise challenge of a gap in credit markets that the private sector is unwilling or unable to fill. Often the private sector is unwilling to lend to lower-income economies. Or the interest rates are just prohibitive to compensate for credit risk, and so transactions don't occur. These agencies were set up many decades ago to provide guarantees to facilitate trade.

CORINNE: Shining a light on these agencies is already helpful. There is a huge role for activism here. Every export credit agency needs to be held to account on financed emissions, and transformed so they are focused on financing emission reductions in low-income economies. Activists in Australia, for example, revealed a huge increase in the funding of fossil fuel projects by Australia's export credit agency.[34] We need to reverse this rapidly. These are financial institutions financed by the public, they make loans which typically generate profits, they support our export sectors, and they provide finance to countries which need it. We need to expand their financing capacity and refocus them on to carbon-reducing investments.

ERIC: If we really want to help with a rapid decarbonization of economies such as India and South Africa, we should raise funding of export credit agencies to target annual disbursements of around 5 per cent of GDP annually. This budget would be solely to fund investments targeting emissions reduction in low-income economies. This is not aid. The aim would be a modest positive financial return and an expansion of green capital goods exports.

CORINNE: A Green Bretton Woods and supercharging bilateral financial institutions will only work for the category of lower-income economies which are already motivated to reduce emissions – such as South Africa and India. What about the rogue states, either deniers, such as Saudi Arabia, or the uncooperative, like Russia.[35] Unfortunately, their emissions are non-trivial.[36]

ERIC: I remain hopeful that through time many of these economies will see the writing on the wall and make progress. Their national industries are in terminal decline. They should really be at the forefront of building capacity in alternatives. In the interim, the smartest approach to these states may be a series of sectoral trade agreements – the GTAs we discussed earlier. To recap, most international trade agreements are not fit for purpose when thinking about climate change. They lack focus and take many, many years to agree, because they aim to cover all sectors of the economy. Viewed through the lens of emissions, we could make far faster progress with sectoral trade agreements for specific industries, such as steel. In particular, this would bring countries like Russia and Brazil into the fold. The objective is straightforward, to accelerate a reduction in emissions in the global steel industry, and get the majority of global steel production to be green by 2035. This is highly ambitious, but the technologies exist. What makes no sense is that Europe aims to fully green its domestic steel production, only for other countries to expand production. As Dieter Helm argues, this amounts to offshoring emissions in the face of a global challenge. We need to establish a set of global rules of the game if any producer in any country wants to participate in the global steel industry. We also need to do the same for cement.

CORINNE: Why is it more effective to narrow trade agreements to the sectoral level?

ERIC: There are a couple of major tactical advantages to this. Firstly, the agreements can be much more focused. Dealing with one sector takes far less time than dealing with ten. And in this instance the priority is determined by contribution to global emissions. A small number of economies dominate the import of steel. If these importers can agree to an ambitious trajectory for green steel standards, the global market would change. Under our proposals, the producing countries would agree a set of EPICs to meet these standards. This is a novel approach to international trade. Less than ten countries dominate global steel and cement production, and international trade accounts for large shares of output. The threat of exclusion from the global market carries teeth. So rather than propose economic sanctions on Russia for its failure to reduce emissions, we can make a global trade agreement for steel which makes access to global markets contingent on steady emissions reductions and compliance with high standards of production. No one can be accused of singling out a given nation – if you meet the standards, you get market access.

CORINNE: The irony, of course, is that countries like Brazil, Russia and Saudi Arabia are all deeply-dependent on the fossil fuel economy, and in Saudi Arabia's and Brazil's case, highly vulnerable to the effects of climate change. By trying to bring them inside a discussion on the emissions of a given sector a gradual change in mindset might occur.

ERIC: Sectoral trade agreements should be negotiated by the

largest importers – in this instance, the EU and the United States.[37] They should set the standards which need to be met. If exporters don't meet them, they get excluded from the global steel market.

One encouraging dimension of the world is the fact that we have more people and more capital engaged in research and development to solve climate change. The three largest economic areas of the world, accounting for 90 per cent of global economic resources, Europe, the United States and China are all focused on this problem. In the context of history, we have never had this amount of intellectual firepower devoted to trying to solve the same challenge. China is critical to this process. It will not surprise people to hear that China accounts for 30 per cent of global emissions, and due in part to its rapid economic growth, its share is rising. What may be more surprising, is that political leadership in China is now highly focused on climate change.[38]

CORINNE: It is worth examining China's motivations, and where it is making progress. It goes without saying that China is not going to significantly forgo economic growth to reduce emissions. The ethical arguments are barely worth rehearsing, but countries like China and India would point to the contribution to cumulative emissions of the United States and Europe.[39] The United States has emitted twice the cumulative levels of greenhouse gases as China, and this would be even more extreme on a per capita basis.

ERIC: That's right. Significantly reducing growth is not a serious option for China. But subject to this constraint, China appears determined to reduce emissions as rapidly as it can, not just to mitigate climate change, but also for a series of important strategic reasons. Like India, it seeks self-sufficiency in energy. Ironically, nationalism aligns with alternative energy in economies which are not self-sufficient in fossil fuels. China imports 40 per cent of its natural gas requirements, and is now the world's largest net importer of oil.[40] Including coal, around 20 per cent of China's energy needs are imported.[41] China feels even more intently the political risk of being dependent on oil, gas and coal imports given the potential for hostility from the United States or its allies. This provides a huge motivation to China to develop domestic resources of wind, solar and nuclear. Two other motivations are central to Chinese strategic policy-making: primacy in emissions-reducing technology and production, and air quality.

CORINNE: How would you characterize China's strengths and weaknesses?

ERIC: China is already a global leader in green technologies.[42] Most of this has occurred through extremely effective policies, utilizing EPICs and regulations to rapidly expand electric vehicles, battery technology, on- and off-shore wind power, thermal and nuclear. The exception is the solar industry. The speed with which China has reached a dominant global share in solar production and facilitated a collapse in prices would be laudable were it not for the fact that it has crowded out lower-emission production, relying on cheap electricity from coal in the manufacturing process, and allegations of appalling labour practices.[43]

CORINNE: Perhaps more than any major economic region,

China has displayed an ability to set ambitious economic plans and deliver on them. It has a unique economic structure, which is directed by the state, but implemented through a combination of state-owned enterprises and a dynamic market-based private sector. This is also evident in its approach to climate change. Through concerted investment in renewables, China has reduced its carbon intensity per unit of GDP by 18 per cent in the five years to 2020, and plans a further 18 per cent reduction by 2025.[44]

ERIC: It is striking that China has sequenced a combination of EPICs and regulations to meet these objectives.[45] Now that electric vehicles are competitively priced and produced at scale in China, the government has targeted a complete phase out of sales of fossil-fueled cars by 2035. They plan for half of their on-road vehicles to be electric or fuel-cell powered, and the other half to be hybrid.[46]

CORINNE: So, what about China's Achilles' heel – coal?

ERIC: Coal dominates China's energy system currently, providing 55 per cent of energy needs, and contributing 75 per cent to overall emissions, given coal is more emissions-intensive than other energy sources. Approximately 25 per cent of the world's coal usage goes to Chinese energy.[47] In contrast to much of the commentary, China is attempting to phase out its dependence on coal, and may well move faster.[48] Although China continues to build new coal-fired electricity generating capacity, it is simultaneously closing less-efficient and worse-polluting capacity. Despite continued rapid economic growth, lower capacity utilization and increased efficiency have caused China's coal consumption to stabilize close to 2013 levels, in notable contrast to India. China can't scale renewables fast enough, in a reliable enough way, to meet economic growth, so coal is the default to meet peak demand.[49]

CORINNE: I would emphasize that it is naive and parochial to think of us "influencing" China.[50] At the margin, we can set standards for the carbon footprint of imports and close access to European and US markets for emissions-intensive Chinese production in areas such as solar or steel. To some extent we will be pushing on an open door. A major difference between China and other emerging economies is that China is not just financially independent, but appears to have excess financial resources. It is financing the largest infrastructure programme in history, the Belt and Road Initiative, which it is increasingly reorienting towards renewables.[51] Deploying a stick approach to influence China is largely a waste of time. Our most powerful strategy is probably to move far faster ourselves. China is competitive. If we transform our economies faster than they are, we may influence their policy.

ERIC: Let's conclude with a quick summary of our proposals for supercharging the world.

CORINNE: We are operating in a context of resurgent nationalist and isolationist tendencies. In advocating for policies to supercharge the world, we can't deny these facts. We segment the world into four categories: high-income economies, motivated low-income economies, rogue states, and China, which is a category of its own. In our previous discussion on supercharging the nation, we outlined how the largest economic blocs in the world – Europe, the United States and China – could accelerate their transitions to sustainability by supercharging investment in sustainable infrastructure and unleashing EPICs in favour of low-emission options across all sectors of their domestic economies. The hack is to exploit a historically low cost of capital to create green assets with a positive return. Not doing so amounts to self-sabotage.

ERIC: The challenge we have considered in our current discussion is to show how high-income or capital-rich economies, like Scandinavia or Germany or China, can accelerate sustainable investment in capital-dependent parts of the world, like India. Relying on acts of generosity or global solidarity doesn't cut it. The policies we advocate result in a much lower cost of capital for low-income economies, rapid growth in infrastructure, and technological transfer. One of the many positive side-effects of sustainable electricity, self-sufficiency, also plays to neo-nationalist instincts. There is an upside for the high-income world, too. Providers of capital get to invest in assets with a positive return, and their capital goods exporters will experience a boom, as will the leaders in green technologies.

CORINNE: We have outlined how existing bilateral agencies, such as export credit agencies, can be supercharged – they need large increases in funding and a refocus on sustainable electricity generation. Activists need to turn their attention to these institutions, which too often operate under the radar, despite their potentially game-changing financial resources. None of this requires new institutions or bureaucracy. Similarly, our advocacy of a Green Bretton Woods repurposes existing financial resources originally ear-marked for post-pandemic liquidity, to be harnessed for financing a global sustainable electricity drive from the IMF. These policies meet the golden rules of good policy: simple, effective and non-partisan. This framework can fit either the ideology of free-marketeers or statists, because it harnesses the best of both. We can tackle climate from the perspective of wealth creation, while addressing moral retribution for those who see the Global North owing a dividend to the Global South, for getting us into this mess in the first place.

ERIC: So far so good, but a strategy is also required for the un-cooperative emitters.

CORINNE: Sanctions are a very blunt tool. There are too many incentives to free-ride, and the target countries often become more entrenched in their views. We propose sectoral trade agreements as a far more productive means to engineering global emission reductions and bring the offenders inside the tent. There is a growing recognition of the need for international cooperation on a sector-by-sector basis. We propose that the WTO be tasked urgently with two global sectoral trade agreements – one for steel and one for cement. Importantly, we are not seeking unanimity among the relevant nations – that would be a sign of failure. We need a global set of standards which put the global sector on a net zero trajectory over 10–15 years. Offenders will be ejected from global trade. Importantly, this is a structure which is not setting out to exclude any individual country. The objective is to set global rules of the game. If you don't play by the rules, you don't get to play.

7

Why this is not a
book about trees

ERIC: "Planting trees could be the number one super-weapon in tackling climate change." This was the headline-grabbing summary of a research paper published in the journal *Science* recently.[1] Greta Thunberg put the conclusions in context: we need trees, but we really need to keep fossil fuels in the ground.[2] Trees and methane are two notable omissions from our discussion so far. Let's start with trees, and then take on cows. So far, there is a clear structure to our argument: work out how significant a contribution every sector is to man-made emissions, assess how tractable it is to solutions, and then apply EPICs, smart regulations or other smart policies. So, what contribution do forests make to global emissions?

CORINNE: Everything we have argued for so far involves reducing how much CO_2 we emit into the atmosphere every year – restricting sources of carbon. There is a separate class of factors which can absorb CO_2 from the atmosphere. These are natural "carbon sinks", such as trees, soil and oceans, and man-made technologies, such as carbon capture and storage. Forests sit in this category. Everyone knows we need to stop the deforestation of the tropics. If you have ever purchased a "carbon offset" when you book a flight, this may well have involved planting trees.

ERIC: How should we think about nature-based carbon sinks compared to all the emissions we are trying to abate?

CORINNE: It is very hard to measure the effect of carbon sinks with accuracy, so the range of estimates is very wide. Oceans are estimated to capture between 25–30 per cent of annual emissions.[3] Forests absorb somewhere between 5 and 15 per cent of man-made emissions.[4] To put both in context, the steel industry alone generates 8 per cent of global emissions. So Greta's summary is correct. Trees are good, but we cannot rely on capture, we need to keep the fossil fuels in the ground.

ERIC: If forests absorb so much less than oceans, why do we hear so much more about planting trees than protecting the oceans?

CORINNE: There is a huge amount we can and should be doing to protect oceans: eradicating plastic and microplastic pollution, heavy metal pollution, overfishing. These are extremely important to the health of the oceans and the preservation of biodiversity, but the ocean's specific ability to absorb carbon emissions is largely out of our control. This is less true of forests, where we can impact their capacity for absorption.

ERIC: What should we do?

CORINNE: The first challenge is to stop deforestation. Around 50 per cent of the world's forests are found in only five countries, Russia, Brazil, Canada, the United States and China, essentially reflecting landmass. The world's forest declines around 0.25 per cent per year, which may not sound like much, but it adds up. Around 95 per cent of this occurs in tropical forests, which have higher biodiversity and carbon sequestration capacity than the northern boreal forests.[5] As with protecting the oceans, there are a number of reasons for protecting our forests. Deforestation causes massive biodiversity loss. Deforestation displaces indigenous peoples and compromises their ecosystem. It causes desertification and aridity. And avoiding deforestation supports one of the earth's major carbon sinks. There is a strong case for action.

ERIC: How tractable is the challenge?

CORINNE: There are two main types of forest loss. One is intentional deforestation, the razing of forests to turn the land to revenue-generating commodity agriculture, such as soy. The other is wildfires, which we have seen on the rise in both tropical and boreal regions. We have managed to slow the rates of net deforestation, but we seem to be better at "afforesting" – creating forests where they previously did not exist – and reforesting, than stopping deforestation in the first place.[6] This is an example of how vested economic interests create large system inefficiencies.

ERIC: Are we on track to stop deforestation, or get net positive?

CORINNE: Some studies suggest we may increase forested land by 10 per cent by 2050. But the two main culprits, razing for agriculture and wildfires, are hard to influence, let alone to supercharge. The stakeholders are fragmented, regimes are not particularly cooperative, particularly Brazil under President Bolsonaro, who is running the Amazon in reverse.

ERIC: How so?

CORINNE: In 2020, NASA revealed that the Amazon rainforest has gone from being a carbon sink to a carbon source, due to the scale of fires from deforestation, as well as raising overall temperatures and drought. This means instead of helping climate change, it's making it worse.[7]

ERIC: This is quite shocking. Bolsonaro has rendered Brazil a rogue state.

CORINNE: The international community is trying to hold Brazil to account, and hopefully matters will improve post-Bolsonaro.[8] The bottom line is that we can't rely on the net forest landscape as part of our pathway to net zero.[9] Partly, because the maths is simply too uncertain. We have a much tighter confidence interval on sources of emissions than we do on carbon sinks, and the levers to influence sinks are hard to grasp. Can we apply EPICs? We can and should try.[10] But we need relevant governments on side, even if just to be trustworthy beneficiaries of incentives to protect forests in their territories.

ERIC: So, what's your conclusion?

CORINNE: We tend to assume forests are carbon sinks, but the Amazon reveals a deeper truth. They can be both carbon sinks and carbon sources, based on the conditions. Sadly, our forests are increasingly becoming carbon sources because of the rates of human deforestation and wildfires. Recent research suggests large parts of the North American boreal forest has become a carbon source, due to wildfires.[11] Critically, this shift is also related to climate change itself. Drought and food scarcity in the tropics further incentivizes deforestation in order to build crops. Hotter, drier conditions in all regions are increasing the incidence and magnitude of fires. The boreal forest fires are one of the famous climate "tipping points": a threshold which, once crossed, leads to large and irreversible systemic changes.

ERIC: What does all this mean for supercharging?

CORINNE: What it tells us is the case for supercharging is stronger than ever. Every month counts. We are well into the territory of risking irreversible "tipping points", or feedback loops. We have to do everything in our power to stop warming as soon as possible. What it also tells us is that how our forest capacity will evolve over the coming years is relatively unknown. Everything in climate science is imprecise, but the confidence intervals are wider here. Because of how these natural ecosystems of oceans and forests are truly that – systems, embedded in a multitude of networks of cause-and-effect. This is not to say we shouldn't be protecting our forests, and doing everything we can to reforest, because there are a multitude of benefits. Trees are good. I think you have a t-shirt that says that, right? They protect biodiversity, soil and water tables. Reforesting creates jobs. They will contribute to the decarbonization journey, in some way. But we just can't rely on them in the "net zero maths" to the extent we currently do.

ERIC: Talk to me about the maths.

CORINNE: All of the scenarios to stay on track for a 1.5 or 2°C scenario assume some degree of "carbon removal", or ways of taking carbon out of the atmosphere.[12] Existing methods, including planting trees, are highly unreliable. This is Greta's point, we should plant trees, but we can't rely on it, or let it distract us from collapsing our greenhouse gas emissions as quickly as possible. That is the first and most important task.

ERIC: You have explained why we can't rely on trees. What about technological solutions for carbon removal?

CORINNE: Trees, soil and the oceans are nature-based solutions to protect or rebuild the world's natural carbon sinks. Carbon capture and storage (CCS) are man-made solutions, which present both as mini-Musk and simple maths problems. Most of the world's largest CCS projects are run by US oil companies, using CO_2 to improve the productivity of oil extraction, which is hardly ideal.[13] There are also lots of highly creative and innovative attempts in many parts of the world to capture and use CO_2 to positive ends.[14] I suspect that over time we will get some very helpful, scalable technologies.

ERIC: One reason you and I have focused on directly targeting emissions reduction through sustainable electricity generation and electrification, is that we already have scalable technology. Our proposals accelerate decarbonization and create wealth, using the state's balance sheet to lower the effective cost of debt to fund the investment. Carbon capture in many ways is the opposite: instead of creating new assets of value which are emissions-free, we keep carbon-emitting assets alive, and then pay for the emissions to be captured, creating a liability for the state. The unfortunate bottom line is that very few viable carbon capture models generate revenues of their own.[15] By contrast, wind power, solar, electric vehicles and new technologies in manufacturing, agriculture and the food chain, can all be profitable activities at the expense of polluting alternatives.

CORINNE: This is why we haven't spent too much time discussing carbon capture. Most capture technologies are effective if they are close to a process of high emissions intensity, and so run the risk of prolonging the life of polluting assets, when our resources should be devoted to closing these assets down. At the same time, we do need CCS because we are unlikely to hit the required emissions reductions targets. My view is largely about tactics. Our proposal for a contingent carbon tax is a powerful incentive to motivate continued innovation in CCS. The problem with many competing proposals to make CCS economically viable is that they usually involve the government paying the private sector for capture and storage.

ERIC: There are risks with the incentives this creates. If a business is based on capturing emissions there is an incentive to produce emissions. An advantage of our proposal for a contingent carbon tax is that the incentive is simply to reduce emissions relative to best practice – either through efficiencies or capture – and no one pays you for the privilege. If a business wants to avoid a windfall tax it needs to be more carbon efficient than its competitors. I prefer this incentive to constantly compete at reducing emissions.

CORINNE: CCS schemes funded by the state also risk becoming complex routes for vested interests to game the system, with the potential for many unintended consequences. In the design of any policy you need to think about what might plausibly go wrong. With most of the policies we advocate, the risk is that they are too successful.

ERIC: Explain?

CORINNE: EPICs often result in a greater than anticipated response – both by consumers and businesses.[16] On the supply side, you can get excess capacity, and on the demand side you get such a large consumer response that the government pays out more than planned. However, from an emissions perspective this is good news. Let's say a boom in wind and solar results in a collapse in electricity prices, companies may go bust. That's not great for the businesses, but it's great for the environment and consumers. What happens if we offer EPICs for electric vehicles and heat pumps and consumers rapidly change behaviour? The government will lose more revenue than planned, but emissions will decline at greater speed. Raising other taxes to cover the losses will be a measure of success. These are good problems to have. Now what could go wrong with a highly successful capture and storage scheme? The worst case is we pay a fortune for capture and emissions don't fall. Why? Because we haven't targeted the source of the emissions, we have just paid for capture. Reducing source emissions has strategic advantages.[17] That's not to say we shouldn't be investing in CCS, or subsidizing it, but just that it's a second-order priority.

ERIC: Removing CO_2 from the atmosphere through natural means, such as oceans and trees, or man-made capture technologies are important, but no substitute for a concerted strategy aimed at emissions reduction. So far, we have focused on CO_2 emissions reduction. How serious a problem is methane?

CORINNE: Anthropogenic methane accounts for roughly 17 per cent of greenhouse gas emissions, measured as "CO_2 equivalents", but is estimated to be responsible for 23 per cent of warming to date.[18] Livestock accounts for roughly 27 per cent of man-made methane emissions. They are one of three major sources of man-made methane, alongside fugitive emissions, or gas leaks, from the mining industry, largely from oil, coal and gas mining, and landfills which happen to emit methane as the contents degrade.[19] Here, there is a strong case for tough regulation. It's difficult to construct positive incentives to stop oil and gas companies from being cavalier about gas leaks. We need to tighten existing regulations, and have zero tolerance and high penalties for rule-breakers.

ERIC: This is a very important point. We have not discussed regulations a great deal so it's important to be clear why. Regulations are a critical leg of any coherent climate strategy, and are currently dominant, but they are most effective in very specific domains and against a backdrop of choice. For example, you won't succeed in banning petrol-driven cars unless you have an electric alternative and a charging infrastructure in place. Regulations are more effective if they are sequenced. EPICs change the economics and build support. They also create an important political constituency which is at odds with the fossil fuel industry.

CORINNE: It is also important to remember that regulations have to be enforced. In many jurisdictions we need stronger regulation, but in many cases regulation simply isn't enforced, which is why closing the global fossil fuel industry and building a sustainable alternative is so important.[20]

ERIC: What about landfills? The thought of managing them is a bit overwhelming. They are huge, sprawling, everywhere, and in the open air? How on earth do you solve their emissions?

CORINNE: The methane is produced by the decomposition of organic waste, such as food, garden waste, paper and wood. The main intervention to date has been to separate this organic waste from household appliances, textiles and other kinds of rubbish and redirect it to specialist facilities that can capture and contain the methane. Project Drawdown outlines the potential for turning the methane into electricity. This model is proven in some locations, and is relatively affordable. Again, it's a case of hard regulations. We need every country to mandate it, and we need to name and shame countries who are not mandating it or not enforcing it. These landfill measures are relatively low cost. It's just a question of shifting the global norm very quickly, such that failing to regulate landfills on their methane management is unacceptable.

ERIC: Back to cows. What do we do about them? We can't ban cows.

CORINNE: Methane from agriculture is the other major man-made source. This is 75 per cent due to animals and 25 per cent rice-paddies, which like wetlands, produce methane because the water blocks oxygen from reaching the soil. Under anaerobic, or oxygen-free conditions, bacteria tend to produce methane. Cows are also responsible for significant nitrous oxide emissions, another major greenhouse gas, as a by-product of their urine. We can't ban cows or rice. It's an area where we need to continue to incentivize market forces and individual behaviour to shift. It's both a mini-Musk and a sheep herding problem. The mini-Musk dimension is well underway. There is extraordinary innovation in plant-based meat and dairy products, which are becoming increasingly close, or preferred, substitutes. Lab-grown dairy is already cost-effective and available, it may not be long before the same is true of meat.

ERIC: How can we accelerate this transition?

CORINNE: Despite the amazing innovation in plant-based substitutes, we are nowhere near the levels of market penetration necessary to use blunt regulations, particularly when we view the problem globally. We have to focus first on the herding sheep tactics we outlined when discussing supercharging the individual. There are stages of mass behavioural change. We know that humans respond to convenience and incentives. If a much cheaper, indistinguishable burger is readily available, it will be the preferred choice. So change starts with EPICs aimed at the relative price of substitutes. When enough people change their behaviour – usually a minority perceived to be greater in number than it really is – the sheep follow. This is reinforced by a change in norms, the emergence of stigma and the introduction of politically-acceptable regulations. In this context there are reasons to be quite optimistic. Plant-based milks are typically substantially more expensive than dairy alternatives and yet have still reached very high penetration rates in the developed world, for ethical and health reasons.[21] Meat and dairy should be the primary targets, there is no way we can get the world to stop eating rice, nor is it close to being a priority.

ERIC: How would you apply EPICs to the meat and dairy industry, and can ethical stigma play an independent role?

CORINNE: Some mass changes, such as the growth in veganism have accelerated due to innovation creating convenience and choice. But as we discussed in the context of supercharging the individual, many people want to live sustainably. The thrust of policy has to be on supporting innovation in this area – we need more Musks – but also in targeting the relative price of substitutes very aggressively, as we have argued consistently. We need innovation to create substitutes for meat and dairy, and we need to deploy all the tools of fiscal policy to shift the relative price in favour of the sustainable choice. At the same time, we should provide EPICs to the agricultural sector to change, as land gets repurposed from livestock and dairy production to alternative uses, whether it is crops for plant-based alternatives, solar farms, or tree-planting. The role of policy is to make these choices financially attractive, not by destroying livelihoods but by creating better alternatives.

ERIC: Ok, so trees, cows and capture, all matter, but not as much as sustainable electricity and getting transport, buildings and industry fully electrified. There are great attempts occurring globally to protect forests, such as the Bonn initiative, and no question, we should be looking at very tough sanctions against rogue states if change doesn't happen fast. Innovation is rapidly emerging to develop scalable alternatives to the livestock and dairy industries. There are already notable successes. The policy objective is clear. We need EPICs to change industry and consumer choice.

Conclusion:
speaking with one voice

It is accepted on every continent that we must transition the global economy to net zero. Other than widespread target-setting, there is not much coherence in global policy-making. Every time we pick up a book on climate change we go straight to the section on what to do about it. Our typical reaction is disappointment. Recommendations range from banal suggestions that we should "find our individual purpose" to implausible ambitions for a complex global carbon tax. A frustration with these proposals helped to motivate us.

This book is all about what to do. We don't need to convince anyone of the urgency. All the major economies of the world are trying to reduce emissions. There is also a clear and emerging consensus on what needs to be done. It doesn't involve a complete overhaul of every individual's lifestyle. It doesn't involve the end of capitalism, reductions in incomes, or huge taxes. We need to start by making two things happen quickly: make electricity sustainable, and run everything off electricity. Everyone *should* know this, and help make it happen.

Our critique is directed at a lack of realism in the current tactics. Realism about human psychology, business and politics. Human beings only change behaviour if their options are cheaper, more convenient and socially normalized. Businesses only change behaviour if they make more money, or if it's made illegal. Policies should follow the golden rules: effective, simple and non-partisan. Far too much energy in climate mitigation is going into efforts that misconstrue these realities.

Supercharge Me is a manual to rapidly collapse emissions. Currently, we are not investing fast enough to halve emissions by 2030 and prevent an increase in global temperatures above 1.5°C. If the policies we describe are implemented, this goal is within range, and shorter timelines are realistic. We are not saying it is easy to electrify everything. It requires huge global investment to overhaul the capital base of the global economy. What we are saying is that it is very doable and already happening, in some cases quite quickly. We have most of the technology we need.

An investment boom on this scale is very likely to make most of the global population better off, and the states that fund it will be wealthier. Given the scale of the threat, and the vast ancillary benefits, we need to move much faster. Charlie Munger says, "Never, ever, think about something else when you should be thinking about the power of incentives".[1] *Supercharge Me* draws on the evidence of the great successes of the last 20 years, in wind power, solar energy and electric vehicles. If we want to collapse the cost of sustainable technologies or change our spending patterns and behaviour, we need to deploy extreme positive incentives for change, or EPICs.

The greatest success stories to date in tackling climate change validate this perspective. They play to our paleolithic emotions: by either making "good actions" very attractive, or bad actions very difficult. We need to set the incentives right for businesses and consumers, and the magnitude of incentives matters. The serendipitous gift of historically low interest rates enables governments to do this without compromising the nation's balance sheet. We have demonstrated ways EPICs can, or already are, being deployed. The evidence is compelling. In contrast to any other policy we know of, the biggest risk with EPICs is that behaviour changes more rapidly than you expect.

This framing – that we have the technology to achieve this radical reduction in emissions, and that EPICs and pragmatic policies are key to scaling them – leads us to a number of concrete policy recommendations. We have sought to shine a light on the interventions that seem to be most effective, and show how they can be scaled.

The worlds of business and finance have undergone a huge cultural shift, and now see the energy transition as inevitable. We need key policy interventions to heavily tilt the playing field in favour of truly sustainable businesses. Rather than wait for carbon pricing at the levels required to deliver impact, we recommend a contingent carbon tax, where governments put the corporate sector on notice, with the risk of emissions turning into a liability on their balance sheet. This incentivizes a race to the top. We recommend green trade agreements (GTAs), starting with the steel sector, where the world's largest importing nations draw up an ambitious timeline and stringent standards to phase out carbon-intensive steel and create a global drive to produce green steel at scale. At the national level we need to reorientate fiscal and monetary policy to accelerate creative destruction. Green mortgages and targeted lending schemes are examples of policies that can be supercharged to have a rapid impact. At the international level, we call for repurposing the vast Covid-19 recovery financial resources to harness the developed world's preferential financing rates for investment in lower-income economies, in a Green Bretton Woods.

Mission critical is to rapidly accelerate investment in sustainable electricity. We know how to do this. We must de-risk investment by providing price floors for investors and collapse the cost of capital, through the banking sector, through government guarantees, and through targeting monetary stimulus at energy infrastructure. The *quid pro quo* is that we require the private sector, which financial markets have primed, to respond with much more ambitious investment timelines.

So, the to-do list at the end of this book depends on who you are. For the policy-makers in finance ministries and governments across the world: harness the state's balance sheet. This demands a refocusing of monetary and fiscal policy. We have explained how to do so, in detail. As rapid investment brings electricity generation as close to 100 per cent sustainable as possible, we need to electrify transport, homes and industry. The sustainable choice has to be much cheaper. Electric vehicles need to be priced well below their petrol-fuelled alternatives. We need extreme incentives to install home solar, to replace gas with electricity, to insulate, and to change our shopping and consuming habits. We need to break the association of "green" with "tax". Of course we should aggressively tax the polluting alternative – but only when there is a clear substitute. We need to target *relative* prices in every sector.

So much for the policy wonks. Most of us are sheep. Thankfully, a minority are activists. Minority activists lead. For the activist reading this book, you have never been more powerful. You won the argument. Social media has supercharged you. For the non-activist, don't worry about how much energy your dishwasher uses. The dishwasher manufacturer is already on it. If you actually want to make a difference, become an activist.

Notes

INTRODUCTION: A FAILURE OF THE MIND

1. J. M. Keynes, *The Collected Writings of John Maynard Keynes, Volume IX: Essays in Persuasion* (London: Macmillan, 1972).
2. Two exceptions to this are Nicholas Stern's *Why Are We Waiting? The Logic, Urgency, and Promise of Tackling Climate Change* (Cambridge, MA: MIT Press, 2015) and Marc Jaccard's *The Citizen's Guide to Climate Success* (Cambridge: Cambridge University Press, 2020). Written by two of the world's experts in the economics of climate change, both of them have significantly influenced our thinking. Stern in his belief that there are huge "co-benefits" to tackling climate change, such as better air quality and jobs, and Jaccard in his withering intolerance of policies that work in textbooks but fail in the face of political reality. Despite these important insights, Stern and Jaccard do not go nearly far enough on the policy front. For example, the fact that governments across the developed world can borrow at the lowest level of interest rates in history is mentioned by neither, and yet this is critical to the economics of funding huge investment in our infrastructure.
3. McKinsey estimate we will need to invest $9 trillion per year for the next three decades, which is an extraordinary 60 per cent increase compared to current annual aggregate spending. See McKinsey & Co., "The net-zero transition and what it means for business", 1 November 2021.
4. Despite the substantial reductions in the emissions intensity of GDP, the trajectory of emissions growth is unchanged, as Dieter Helm points out in *Net Zero: How We Stop Causing Climate Change* (London: William Collins, 2020). Emissions intensity, or CO_2 per unit of GDP growth, has decreased, which is at least some progress. Ultimately what matters for the temperature trajectory is absolute emissions.
5. D. Wallace-Wells, *The Uninhabitable Earth* (New York: Allen Lane, 2019). See also B. McGibben, *Falter: Has The Human Game Begun to Play Itself Out?* (New York: Henry Holt, 2019).
6. Including the Intergovernmental Panel on Climate Change (IPCC),

the Energy Transitions Commission (ETC), and the United Nations Framework Convention on Climate Change (UNFCCC).
7. R. Henderson, *Reimagining Capitalism* (London: Penguin, 2020); K. Raworth, *Doughnut Economics* (London: Random House, 2017).
8. California uses a combination of EPICs and regulations; see https://www.gov.ca.gov/2020/09/23/governor-newsom-announces-california-will-phase-out-gasoline-powered-cars-drastically-reduce-demand-for-fossil-fuel-in-californias-fight-against-climate-change/.
9. E. Lonergan and M. Blyth, *Angrynomics* (Newcastle upon Tyne: Agenda, 2020).
10. Mark Jaccard (2020) has heavily influenced our thinking here. He refers to flexible regulations. We have tried to clarify the litmus test for regulation, which may seem obvious but often is not the primary consideration; *are they simple and do they work?* We are calling these "smart" regulations with a hat-tip to his more nuanced concept.
11. Steel production is responsible for roughly 8 per cent of global emissions, whilst UK emissions roughly 1 per cent. See World Steel Association, "Steel's contribution to a low carbon future and climate resilient societies" (2020) and Our World in Data, "United Kingdom: CO_2 country profile" https://ourworldindata.org/co2/country/united-kingdom?country=-GBR.
12. Climate science is probabilistic. We are operating under great uncertainty in assessing both how atmospheric CO_2 concentrations translate into average temperature rises, and the impact of those temperature rises on climate and ecology. There is strong scientific consensus that in order to reduce the risk of a number of irreversible climate tipping points, we need to keep temperatures below 1.5°C warmer than 1900. This is called the 1.5°C scenario. It is estimated that we have a "budget" of 400–600 gigatons of CO_2 equivalent emissions to keep within 1.5°C. Given annual emissions are roughly 43 gigatons CO_2e and rising, we have ten years to almost entirely decarbonize our economy. Stern points out that we have already exceeded the CO_2 concentration that modern *Homo sapiens* has experienced, which puts in context the unknowns ahead of us. See D. Luthi *et al.*, "High-resolution carbon dioxide concentration record 650,000–800,000 years before present", *Nature* 453 (2008), 379–82.

1. SUPERCHARGING: WHAT IS IT?

1. Achieving this requires halving emissions in the next decade. According to the IPCC 1.5°C scenario, we need to reduce emissions to 25–30 GtCO$_2$e per year, down from between 52 GtCO$_2$, or 57 GtCO$_2$, including changes in land use. See H. Ritchie and

M. Roser, "CO_2 and greenhouse gas emissions" (2020), Our World in Data, https://ourworldindata.org/co2-and-other-greenhouse-gas-emissions.

2. The emissions *intensity* of the world economy has fallen sharply, but not the level of emissions. Emissions intensity is the volume of GHG emissions per unit of GDP growth. A decrease in emissions indicates we are "decoupling" GHG emissions from economic growth. According to the PWC Net Zero Economy Index, every member of the G20 has reduced emissions intensity since the 2015 Paris Agreement. However, the global rate of progress, 2.4 per cent and 2.5 per cent in 2019 and 2020 respectively is a long way off the 12.9 per cent average annual emission intensity reduction required to align with the 1.5°C scenario; see PricewaterhouseCoopers, *Net Zero Economy Index 2021*; https://www.pwc.co.uk/services/sustainability-climate-change/insights/net-zero-economy-index.html. The alignment gap between targets and policies is tracked by Carbon Action Tracker, and explored in a 2021 *Nature* study: "... we find that the probabilities of meeting their nationally determined contributions for the largest emitters are low, e.g. 2% for the USA and 16% for China. On current trends, the probability of staying below 2°C of warming is only 5%, but if all countries meet their nationally determined contributions and continue to reduce emissions at the same rate after 2030, it rises to 26%". Climate Action Tracker, "2100 warming projections: emissions and expected warming based on current pledges and policies" (last updated July 2021), https://climateactiontracker.org/global/temperatures/; P. Liu and A. Raftery, "Country-based rate of emissions reductions should increase by 80% beyond nationally determined contributions to meet the 2°C target", *Communications Earth and Environment* 2(29) (2021), https://doi.org/10.1038/s43247-021-00097-8.

3. The Nobel Prize-winning economist, Thomas Schelling, in his prescient set of essays on climate change written almost 20 years ago, highlights the relative success of strategies aimed at actions rather than results. See T. Schelling, *Strategies of Commitment and Other Essays* (Cambridge, MA: Harvard University Press, 2006), 40.

4. There is enormous variance by regions and cities across these geographies. For the differential performance across US states, see D. Saha and J. Jaeger, "Ranking 41 US States Decoupling Emissions and GDP Growth", World Resources Institute, https://www.wri.org/insights/ranking-41-us-states-decoupling-emissions-and-gdp-growth. For variance across global cities, see CNCA, https://carbonneutralcities.org/cities/.

5. Some of the best work on this has been done by the economist Carlota Perez; see C. Perez, *Technological Revolutions and Financial*

Capital (Cheltenham: Elgar, 2002). The point is also emphasized in Stern, *Why Are We Waiting?*

6. According to estimates in the *Financial Times*, the losses to the oil and gas sector could exceed $900 billion. As a percentage of the value of total global equity assets, it is relatively small (see "The $900bn cost of stranded energy assets", *Financial Times*, 4 February 2020). For a list of the world's largest oil producers, see https://www.worldometers.info/oil/oil-production-by-country/. The oil sector's market capitalization has been falling significantly over the last decade, and the energy sector only accounts for around 2 per cent of the S&P500, and 3.5 per cent of the MSCI World. Among the major asset classes equities are the most unequally distributed, and oil is a relatively small share of the total value. For more on thugocracies, see N. Ries, "Thugocracy: bandit regimes and state capture", *Safundi* 21:4 (2020), 473–85.

7. There are genuine cases where humanitarian support may be warranted, such as Nigeria where 90 per cent of export earnings are generated by oil. At a global level, according to the European Commission's science hub, approximately 58 million people are employed in the energy sector, with half of these employed in fossil fuels. This is a relatively small fraction of the global labour force, approximately 0.7 per cent. To be clear, changes in aggregate employment through the cycle are far more significant. The rate of growth in sustainable energy jobs is also significant. There is some evidence this will be higher in a renewables based economy. See J. Jaeger *et al.*, "The green jobs advantage: how climate-friendly investments are better job creators" (October 2021), World Resources Institute; https://www.wri.org/research/green-jobs-advantage-how-climate-friendly-investments-are-better-job-creators, and V. Czako, "Employment in the energy sector", JRC Science for Policy Report, 9 July 2020, https://ec.europa.eu/jrc/en/science-update/employment-energy-sector.

8. D. Ariely, "Human behaviors vs climate crisis (part one)", YouTube, 4 June 2012, https://www.youtube.com/watch?v=mJpgeJagZzQ. The incomprehensible nature of the problem, and consequent challenge of winning "political will" for effective climate policy is also explored in E. Kamarck, "The challenging politics of climate change", Brookings Institution, 23 September 2019, https://www.brookings.edu/research/the-challenging-politics-of-climate-change.

9. Jaccard, *The Citizen's Guide to Climate Success.*

10. B. Gates, *How To Avoid A Climate Disaster* (New York: Random House, 2021).

11. *Ibid.* For emissions data, presented clearly, see Our World in Data,

https://ourworldindata.org/, which is based on CAIT climate data, which is aligned to the IPCC.

12. Electrifying heavy goods vehicles (HVGs) which travel long distances may be an exception to this, although innovation is happening at extraordinary speed in this area. Battery weight is a major consideration, and current low (albeit increasing) battery energy density means that heavy-duty battery electric vehicle (BEV) trucks struggle to be technologically viable and cost competitive with internal combustion engine (ICE) trucks; see ETC, "Making clean electrification possible: 30 years to electrify the global economy" (2021), https://www.energy-transitions.org/wp-content/uploads/2021/04/ETC-Global-Power-Report-.pdf.

13. Through either direct or indirect electrification. Indirect electrification refers to provision of energy through fuels such as hydrogen, ammonia or synfuels which can be produced "cleanly" using renewable electricity. ETC predicts $c.80$ per cent of industry will be electrified (directly or indirectly) by 2050, see "Making mission possible: delivering a net-zero economy" (2020), https://www.energy-transitions.org/wp-content/uploads/2020/09/Making-Mission-Possible-Full-Report.pdf; see also McKinsey & Co., "Plugging in: what electrification can do for industry", https://www.mckinsey.com/industries/electric-power-and-natural-gas/our-insights/plugging-in-what-electrification-can-do-for-industry. The most challenging sectors to electrify are those that require very high temperature heat (>1000°C) or chemical transformation processes that output carbon dioxide. The cement, steel and chemical industries are the most notable examples of industrial/manufacturing sectors with these requirements, with electrification technologies still in varying degrees of development. See ETC, "Mission possible: reaching net-zero carbon emissions from harder-to-abate sectors by mid-century", https://www.energy-transitions.org/publications/mission-possible/.

14. Huge progress is being made with electrifying trucks and heavy vehicles, for example, Tiger Trailers (https://tigertrailers.co.uk/road-transport-hub/technology-electric-hybrid-trucks/).

15. This assumes that agriculture is 15 per cent emissions (not 20 per cent) and less than 10 per cent emissions can be reduced through electrification; aviation is $c.3$ per cent emissions and can't yet be electrified; shipping is $c.3$ per cent emissions and majority can't (yet) be electrified. Electrification (direct and indirect) will only account for $c.80$ per cent of energy needs of industry; 20 per cent of industry energy equates to 6 per cent of overall emissions not electrified. The balance amounts to 25.5 per cent that can't be electrified.

16. Estimates of timelines include 9–18 months for a 50 MW farm,

1–6 months for 0.5 MW farm, 3 years for a 1 GW farm. Site assessment, planning and development can take up to two years, and in general take longer than construction and installation. See Coriolis Energy, "Wind farm life cycle" (2020), http://coriolis-energy.com/landowners/wind_farm_life_cycle.html; Renewables First, "What is a wind turbine project timeline?" (2015), https://www.renewablesfirst.co.uk/windpower/windpower-learning-centre/how-long-will-the-whole-project-take/; BVG Associates, "Guide to an offshore wind farm" (2019), https://www.thecrownestate.co.uk/media/2860/guide-to-offshore-wind-farm-2019.pdf.

17. The precise number depends on a series of factors, principally how much energy the houses use and the utilization rate of the turbines. See US Geological Survey, "How much wind energy does it take to power an average home?", https://www.usgs.gov/faqs/how-much-wind-energy-does-it-take-power-average-home?

18. "Nimbyism" (NIMBY: Not-In-My-BackYard) is a characterization of general support for a development, except when it is a specific local project. The evidence isn't conclusive that nimbyism per se is a major barrier to rapid scaling of renewable capacity. It is most potent when it translates into policy, such as planning regulations. For example, the UK Conservative Party introduced restrictive requirements for onshore wind in 2015, which resulted in a 96 per cent decrease in application submissions and granted permissions. See S. Jarvis, "The economic costs of NIMBYism: evidence from renewable energy projects", Energy Institute at Haas, January 2021, https://haas.berkeley.edu/wp-content/uploads/WP311.pdf; S. Carley et al., "Energy infrastructure, NIMBYism, and public opinion: a systematic literature review of three decades of empirical survey literature", *Environmental Research Letters* 15:9 (2020), 093007, https://doi.org/10.1088/1748-9326/ab875d; and R. Windemer, "Onshore wind farm restrictions continue to stifle Britain's renewable energy potential", The Conversation, 12 October 2020, https://theconversation.com/onshore-wind-farm-restrictions-continue-to-stifle-britains-renewable-energy-potential-147812.

19. Hornsea Two, a 1.4 GW wind farm off the Yorkshire coast, will power 1.3 million homes, with a projected lead time from start to finish of around five years (2.5–3 years' build time); https://hornseaprojects.co.uk/hornsea-project-two/about-the-project#honsea-project-two-timeline-september-2017.

20. For details on UK offshore wind, see Rystad Energy, "UK's renewable energy capacity set to double by 2026, when offshore wind will overtake onshore", 28 October 2020, https://www.rystadenergy.com/newsevents/news/press-releases/uks-renewable-energy-

capacity-set-to-double-by-2026-when-offshore-wind-will-overtake-onshore/

21. The UK's private sector is arguing for more ambitious onshore wind development. See "Industry urges Government to set new target to double UK onshore wind capacity by 2030", RenewableUK, 13 October 2021, https://www.renewableuk.com/news/583055/Industry-urges-Government-to-set-new-target-to-double-UK-onshore-wind-capacity-by-2030.htm.

22. X. Lu *et al.* (2009) "Global potential for wind-generated electricity", *Proceedings of the National Academy of Sciences* 106:27 (2009), 10933–8. This aligns with IEA, *World Energy Outlook 2019*, https://www.iea.org/reports/world-energy-outlook-2019. Further analysis also highlights only 1 per cent of global land area would have to be utilized by solar farms to produce 100,000 TWh of electricity (versus 27,000 TWh of total electricity generated today from all sources), see ETC, "Making clean electrification possible". The US numbers are even more impressive. The US has 15,200 GW of available onshore and offshore wind resources, and total current utility-scale electricity capacity of 1,117 GW; see US Energy Information Administration (EIA), "Electricity generation, capacity, and sales in the United States" (February 2021), https://www.eia.gov/energyexplained/electricity/electricity-in-the-us-generation-capacity-and-sales.php; "U.S. renewable energy factsheet", Center for Sustainable Systems, University of Michigan, Pub. No. CSS03-12 (2021); and M. Shaner *et al.*, "Geophysical constraints on the reliability of solar and wind power in the United States", *Energy & Environmental Science* 11 (2018), 914–25, https://pubs.rsc.org/en/content/articlelanding/2018/ee/c7ee03029k#!divAbstract.

23. J. Abraham, "Wind and solar can power most of the United States", *The Guardian*, 26 March 2018, https://www.theguardian.com/environment/climate-consensus-97-per-cent/2018/mar/26/study-wind-and-solar-can-power-most-of-the-united-states.

24. IRENA, "Renewable energy power capacity country rankings", https://irena.org/Statistics/View-Data-by-Topic/Capacity-and-Generation/Country-Rankings.

25. Of the traditional sources of electricity generation, gas has the shortest lead times, and is closest to renewables, coal and nuclear have much longer lead times; see, e.g., "Sakaka Photovoltaic Solar Project", https://www.power-technology.com/projects/sakaka-photovoltaic-solar-project/ and YSG Solar, "How long does it take to construct a solar farm?", https://www.ysgsolar.com/blog/how-long-does-it-take-construct-solar-farm-ysg-solar.

26. Wind and solar was 91 per cent of new renewables, with total new renewables representing 82 per cent of net electricity generation

capacity expansions (IRENA, "Renewable capacity highlights 2021"). China alone will account for almost half of renewable electricity growth in 2021 – an extraordinary fact in-of-itself, see IEA, *Global Energy Review 2021*.

27. For perspective, the Energy Transitions Commission projects current global solar and wind capacities of 1,300 GW will need to increase to 40,000–51,000 GW by 2050 for 83 per cent of energy needs to be met by direct or indirect renewable electricity. We think this timeline can be shortened significantly.

28. The geographic divergences in performance in the US appear mainly to be political and economic; see "Path to carbon-free power generation by 2035", S&P Global, https://pages.marketintelligence. spglobal.com/rs/565-BDO-100/images/path-to-carbon-free-power-generation-by-2035-infographic.pdf. On the efforts of individual US utilities, see "Path to net-zero", S&P Global.

29. In theory there should be little difference between increasing the tax on fuel-based vehicles relative to electric vehicles in order to create a large relative price differential in favour of EVs, but the behavioural experience and political consequences of achieving the same price differential via a tax exemption on the EV may determine the policy's success or failure.

30. Gates, *How To Avoid A Climate Disaster*, 189–94.

31. D. Hume, *A Treatise on Human Nature*, Book III, Part 2: Of the Origin of Justice and Property [1739].

32. J. Hughes, C. Knittel and D. Sperling, "Evidence of a shift in the short-run price elasticity of gasoline demand", *Energy Journal* 29:1 (2008), 113–34.

33. Economists have focused too much on the microeconomics of externalities, and not enough on the price elasticity of demand. The existence of substitutes is a primary determinant of price elasticity. In order to affect the latter we need to target the creation of substitutes and their relative price. For a definition of price elasticity and its determinants, this primer from the OECD is helpful, https://stats.oecd.org/glossary/detail.asp?ID=3206 (from *Glossary of Industrial Organisation Economics and Competition Law*, compiled by R. Khemani and D. Shapiro).

34. Mark Jaccard argues that those professionally involved in the tobacco industry were less likely to accept that cigarettes cause cancer, even if they accepted scientific evidence in most other aspects of their lives. See Jaccard, *Citizen's Guide to Climate Success*.

35. This approach is supported by recent innovative research by MIT economist Daron Acemoglu. He develops a theoretical growth model with clean and dirty technologies competing in production and innovation – the results of this model highlight optimal policy

relying initially on aggressive research subsidies (i.e. an EPIC) that diminish over time, complemented by back-loaded and increasing carbon taxes; see Acemoglu *et al*, "Transition to clean technology", *Journal of Political Economy* 124:1 (2016), https://doi.org/10.1086/684511.

36. California's policy towards electric vehicles is a classic example of this sequencing, starting with EPICs to create behavioural change and economies of scale, then adopting ever-stricter regulations. For the list of incentives, see "Save money on transportation", Clean Vehicle Rebate Project, https://cleanvehiclerebate.org/eng/ev/benefits/save-money; for planned regulations phasing out ICEs, see Governor Newsom's announcement, https://www.gov.ca.gov/2020/09/23/governor-newsom-announces-california-will-phase-out-gasoline-powered-cars-drastically-reduce-demand-for-fossil-fuel-in-californias-fight-against-climate-change/.

37. See "How governments spurred the rise of solar power", *The Economist*, 7 January 2021, https://www.economist.com/technology-quarterly/2021/01/07/how-governments-spurred-the-rise-of-solar-power. Solar energy has seen the steepest decline in price of any electricity technology over the last decade, a 95 per cent collapse in price; see IRENA, *Renewable Power Generation Costs in 2019*. By 2021, global installed capacity is approaching 800 GW, and growing at 20 per cent per year; see "New Energy Outlook 2021", BloombergNEF, https://about.bnef.com/new-energy-outlook/.

38. See O. Kimura and T. Suzuki, "30 years of solar energy development in Japan: co-evolution process of technology, policies, and the market", Berlin Conference on the Human Dimensions of Global Environmental Change, November 2006, https://citeseerx.ist.psu.edu/viewdoc/download?doi=10.1.1.454.8221&rep=rep1&type=pdf.

39. We discuss China's controversial role in the development of the global solar industry in Chapter 6.

40. Norwegian EPICs are changing over time. The numbers cited here are drawn from the Norwegian Electric Vehicle Association (Norsk Elbilforening), https://elbil.no/english/norwegian-ev-policy/ (accessed August 2021); see also L. Fridstrom *et al.* (2021), "The Norwegian vehicle electrification policy and its implicit price of carbon", *Sustainability* 13:3 (2021), 2071–105 and "Electric cars: why little Norway leads the world in EV usage", *Forbes*, 18 June 2019.

41. For a global comparison of fuel duties, see the US Department of Energy's Alternative Fuels Data Center", https://afdc.energy.gov/data/.

42. See, for example, the Taxpayers' Alliance, "Fuel duty", https://www.taxpayersalliance.com/fuel_duty_briefing.

43. The IEA article, "How global electric car sales defied Covid-19 in

2020" has very interesting data and narratives on EV sales over the pandemic, with heavy focus on how subsidies/support packages boosted uptake in some countries versus others; see https://www. iea.org/commentaries/how-global-electric-car-sales-defied-covid-19-in-2020.

44. See "China's electric car capital has lessons for the rest of the world", *Bloomberg*, 26 June 2021.

45. This is something Martin Wolf and others have written on extensively; see, for example, "Now is the time to reform the UK's dysfunctional tax system", *Financial Times*, 7 February 2021. Taxes on emissions are only one part of sensible aggregate tax policy.

46. This will also require EPICs to encourage more rapid replacement. The average life of a vehicle varies significantly by geography; EU average age of fleet on the road is 11.5 years, and 12 years in the US. Large one-off incentives may need to be deployed to accelerate trade-ins for electric vehicles; see ACEA, "Average age of the EU vehicle fleet, by country", https://www.acea.auto/figure/average-age-of-eu-vehicle-fleet-by-country/.

47. IEA, *Global EV Outlook 2021*, https://www.iea.org/reports/global-ev-outlook-2021/trends-and-developments-in-electric-vehicle-markets.

48. See the OECD report, *Investing in Climate, Investing in Growth* (Paris: OECD, 2017) prepared for Germany's presidency of G20.

49. The Energy Transitions Commission estimates required investments, on average, of $1.2–1.4 trillion per annum until 2050 for widespread wind and solar generation roll-out, versus the $300 billion spent in 2020; see ETC, "Making clean electrification possible".

50. See Stern, *Why Are We Waiting?* and C. Perez, "Digital and green: a very convenient marriage", Beyond the Technological Revolution, 3 November 2020, http://beyondthetechrevolution.com/blog/digital-and-green-a-very-convenient-marriage/.

51. In their most recent economic assessment, the Intergovernmental Panel on Climate Change estimated that the net present value of damages from 2°C of warming (relative to pre-industrial times) would amount to $69 trillion, with this figure rising at a faster rate as temperature projections increase beyond 2°C; see O. Hoegh-Guldberg *et al.*, "Impacts of 1.5°C global warming on natural and human systems" in *Global Warming of 1.5°C: An IPCC Special Report* ... (2018), www.ipcc.ch/site/assets/uploads/sites/2/2019/05/SR15_Chapter3_Low_Res.pdf. See Carlota Perez's seminal thesis on the relationship between technological development and economic booms; C. Perez, *Technological Revolutions and Financial Capital* (Cheltenham: Elgar, 2002).

52. "The $900bn cost of 'stranded energy assets'", *Financial Times*, 4 February 2020. A report by Credit Suisse estimates the global wealth stock at $418 trillion by the end of 2020, by which calculus the *Financial Times* estimates that stranded assets account for 0.2 per cent of global wealth stock; see Credit Suisse Research Institute, *Global Wealth Report 2021*, http://docs.dpaq.de/17706-global-wealth-report-2021-en.pdf.

53. These are our estimates. Corporate assets are a subset of wealth, we would expect other forms of wealth, such as housing to be more evenly distributed. Similarly, financial assets, such as bonds and bank deposits are likely to be widely held, albeit unequally distributed. Stock market wealth and private business ownership, tend to be the most concentrated forms of wealth. See A. Advani, G. Bangham and J. Leslie, "The UK's wealth distribution and characteristics of high-wealth households", Warwick Economics Research Papers No. 1367 (August 2021), https://warwick.ac.uk/fac/soc/economics/research/workingpapers/2021/twerp_1367_-_advani.pdf.

54. Approximately 6 per cent of the global population own shares, and the ownership of shares is highly concentrated; see P. Grout, W. Megginson and A. Zalewska, "One half-billion shareholders and counting: determinants of individual share ownership around the world" (Dec 2009), http://dx.doi.org/10.2139/ssrn.1364765. For the United States, which has the world's largest stock market, see R. Wigglesworth, "How America's 1 per cent came to dominate equity ownership", *Financial Times*, 11 February 2020, https://www.ft.com/content/2501e154-4789-11ea-aeb3-955839e06441.

55. In fact, recent analysis highlights that across most of the sectors that pose the greatest decarbonization challenge, the impact on final product cost for consumers is very low – for example, the price increase of an average car made from "green" steel would only be $180, *representing* a 1 per cent rise in the price versus a car made from carbon intensive steel, see ETC, "Making mission possible".

56. Based on emissions trajectories underpinned by current conditional national commitments, we are set to emit 54 billion tonnes of greenhouse gases in 2030, versus a required maximum emission level of 25 billion tonnes if we are to mitigate temperature rise over 1.5°C; see UNEP, *Emissions Gap Report 2020*, https://www.unep.org/resources/emissions-gap-report-2021.

2. CORPORATE PHILOSOPHERS

1. Financed emissions are those produced by the companies which financial institutions lend to, invest in, or otherwise support.

2. Debate at the Harvard Museum of Natural History, 9 September 2009; quoted in Henderson, *Reimagining Capitalism*.

3. "Firms ignoring climate crisis will go bankrupt, says Mark Carney", *The Guardian*, 13 October 2019, https://www.theguardian.com/ environment/2019/oct/13/firms-ignoring-climate-crisis-bankrupt-mark-carney-bank-england-governor.

4. According to the London School of Economics's Grantham Research Institute on Climate Change, there has been a more than 20-fold increase in climate laws since 1997. Litigation cases are also rising exponentially. See the Climate Change Laws of the World database, https://climate-laws.org.

5. "Carbon border adjustment mechanism: questions and answers", European Commission, 14 July 2021, https://ec.europa.eu/ commission/presscorner/detail/en/qanda_21_3661.

6. Stern goes on to say, ". . . with good policy and strong commitment, the low-carbon transformation can be the real dynamic growth story of the future. It could have still greater potential than previous technological revolutions to improve world living standards and quality of life" (Stern, *Why Are We Waiting?*, 36–7). See also C. Perez and T. Leach, "A smart green 'European way of life': the path for growth, jobs and wellbeing", BTTR working paper, March 2018, http://beyondthetechrevolution.com/wp-content/ uploads/2014/10/BTTR_WP_2018-1.pdf.

7. This is based on a small sample and includes households which consume both dairy milk and plant-based alternatives; see T. Malone, C. Wolf and B. McFadden, "Who is substituting milk with plant-based beverages and why?", College of Agriculture and Natural Resources, Michigan State University, 10 November 2020, https://www.canr.msu.edu/news/who-is-substituting-milk-with-plant-based-beverages-and-why.

8. See C. Driebusch, "Oatly IPO prices at $17 a share, notching $10 billion valuation", *Wall Street Journal*, 19 May 2021, https://www.wsj. com/articles/oatly-prepares-to-price-ipo-amid-stock-market-declines-11621456361 and J. Evans and E. Terazono, "The battle for the future of milk", *Financial Times*, 7 May 2021, https://www.ft.com/ content/da70e996-a70b-484d-b3e6-ea8229253fc4.

9. "Orsted hikes renewables target to 50 GW; US unveils California offshore wind zones", Reuters, 9 June 2021, https://www. reutersevents.com/renewables/wind/orsted-hikes-renewables-target-50-gw-us-unveils-california-offshore-wind-zones. At the time of writing (Q4 2021), Orsted shares were trading at 899 DKK, up from 275 DKK five years prior.

10. The price of shares in Exxon, BP and Shell have fallen in the five years to August 2021, in sharp contrast to Orsted.

11. We are not making an observation on how "sustainable" Oatly is as a business, beyond the observation that plant-based milk is on average less emissions intensive than dairy milk. See J. Poore and T. Nemecek, "Reducing food's environmental impacts through producers and consumers", *Science* 360:6392 (2018), 987–92, https://www.science.org/doi/10.1126/science.aaq0216.

12. Carbon Footprint, "Mandatory greenhouse gas (GHG) reporting", https://www.carbonfootprint.com/mandatorycarbonreporting.html.

13. The Carbon Disclosure Project (CDP) is a not-for-profit that runs a disclosure system for investors, companies, cities, states and regions; see https://www.cdp.net/en/companies/companies-scores.

14. The Carbon Disclosure Project (CDP), UN Global Compact (UNCG), World Resources Institute (WRI) and the World Wildlife Fund for Nature (WWF); see https://sciencebasedtargets.org.

15. As of 2021, SBTi has mandated that these targets have to be in line with the 1.5°C pathway.

16. Under the system of targets, the company needs a base year, a target year for delivery and a percentage decrease in emissions. Walmart's base year is 2015, its targeted decrease in emissions is 18 per cent, and the target year to deliver that change is 2025. When a company "submits" a target, it is then verified by SBTi, to check it is both sufficiently ambitious, but also that the company has a plan to deliver on it.

17. The scopes framework was developed by the Greenhouse Gas Protocol, https://ghgprotocol.org/about-us.

18. The SBTi specifies that if a company's Scope 3 emissions are more than 40 per cent of total, they have to set a Scope 3 reduction target. Moreover, there is a minimal requirement for reduction in Scope 1 and 2 emissions, as they are often easier. Walmart set an 18 per cent reduction target for Scope 1 and 2. For Scope 3, they set an absolute target, rather than relative, to reduce emissions by a billion tons, or a gigaton, by 2030 from a 2015 baseline.

19. We have seen widespread adoption since 2018, when just 216 corporates had signed up. SBTi provides a database of companies who have committed and/or have set targets, including temperature alignment of targets; see https://sciencebasedtargets.org/companies-taking-action.

20. In addition to the SBTi, which is a target-based commitment, there has also been huge progress on the disclosure of climate risks. The international Taskforce on Climate-Related Financial Disclosure (TCFD) is a reporting framework that mandates reporting on climate risk for companies and financial institutions. A number of countries have made it mandatory to report on emissions from own

operations, including the UK and China, and the G7 has proposed making it mandatory. In the US, it is not mandatory, but 25 per cent of the S&P500 now cite climate risk in their annual reports to the SEC. Fund managers and corporate CEOs are now all studying these risk reports which are now mandatory. The perception and understanding of climate risks have changed on the back of these reporting practices, and that will impact how leaders act on them.

21. A "windfall" tax is a one-off tax on profits. "Majority of UK public supports windfall tax", *Financial Times*, 17 May 2020, https://www.ft.com/content/b7441bee-6bf7-46c2-ab75-916fec31f521.

22. The variance in ambition across the world's major utilities is a prime case in point. We need to incentivize a race to the top. See S&P Global, "Path to net-zero".

23. See Jaccard, *Citizen's Guide to Climate Success*, chapter 6.

24. See Richard Rosen, "Carbon taxes: a good idea, but can they be effective?" Institute for New Economic Thinking, 28 June 2021, https://www.ineteconomics.org/perspectives/blog/carbon-taxes-a-good-idea-but-can-they-be-effective and Adam Tooze, "Present at the creation of a climate alliance – or climate conflict", *Foreign Policy*, 6 August 2021.

25. World Steel Association, "Climate change and the production of iron and steel" (2021), https://www.worldsteel.org/en/dam/jcr:228be1e4-5171-4602-b1e3-63df9ed394f5/worldsteel_climatechange_policy%2520paper.pdf.

26. Mission Possible Partnership, "Steeling demand: mobilizing buyers to bring net-zero steel to market before 2030" (2021), www.energy-transitions.org/wp-content/uploads/2021/07/2021-ETC-Steel-demand-Report-Final.pdf.

27. "ArcelorMittal Europe to produce 'green steel' starting in 2020", https://corporate.arcelormittal.com/media/news-articles/arcelormittal-europe-to-produce-green-steel-starting-in-2020.

28. Free-riding occurs where those who benefit from a shared resource, or service, do not pay for it. As people are either not paying their fair share or not at all, it is considered a form of market failure.

29. D. Helm, "A carbon border tax can curb climate change", *Financial Times*, 5 September 2010, https://www.ft.com/content/a68bfc80-b915-11df-99be-00144feabdc0. For details of the EU proposal, see "Carbon border adjustment mechanism", European Commission.

30. See M. Kettunen *et al.*, "An EU green deal for trade policy and the environment: aligning trade with climate and sustainable development objectives", IEEP Brussels, and Tooze, "Present at the creation of a climate alliance – or climate conflict".

31. Our World In Data has a neat infographic showing net emissions imported or exported, which demonstrates Europe's role as a

significant emissions importer; https://ourworldindata.org/consumption-based-co2.

32. The 2021 bilateral EU–US steel agreement shows how this can be done rapidly, and also how carbon standards can be agreed for a single sector. Unfortunately it does not go far enough. The standards need to be tougher, and there needs to be an independent agreement on tariff-exemptions for EPIC green steel. A hat tip to Soumaya Keynes for this suggestion. The policy objective is very simple: make global green steel – in every significant marketplace – cheaper than the carbon-intensive alternative. See U. Dadush, "What to make of the EU-US deal on steel and aluminium?" Bruegel, 4 November 2021, https://www.bruegel.org/2021/11/what-to-make-of-the-eu-us-deal-on-steel-and-aluminium/.

33. There are emerging proposals around industry-specific green trade agreements; see T. Tucker and T. Meyer, "A green steel deal: toward pro-jobs, pro-climate transatlantic cooperation on carbon border measures", The Roosevelt Institute, June 2021, https://rooseveltinstitute.org/wp-content/uploads/2021/06/RI_GreenSteelDeal_WorkingPaper_202106.pdf.

34. There is a vast literature on incentives as motivators. The classic examples include incentives to get children to do homework, or incentivizing employees in the workplace. The conclusions of this literature on positive versus negative incentives is inconclusive, and much hinges on the role of "intrinsic" motivation to act. The kinds of incentives we are talking about are qualitatively different. We assume no hardship, motivation or willpower required. This is not about doing homework instead of watching television, but rather swapping one substitute for another, where often the purchaser or consumer is indifferent to the product aside from the price and quality. Assuming quality is constant, price matters. And people have a different psychological response to a sense of something being a "good deal" versus being "screwed over".

3. MONEY GETS THE MESSAGE

1. K. Löffler, A. Petreski & A. Stephan, "Drivers of green bond issuance and new evidence on the 'greenium'". *Eurasian Economic Review* 11 (2021), 1–24, https://doi.org/10.1007/s40822-020-00165-y.

2. Scores based on ESG criteria are now being given to all companies listed on the stock markets, and to issuers of bonds – a way in which governments and companies raise debt. For more detail on how markets are using ESG criteria see, MSCI, "What is ESG?", https://www.msci.com/esg-101-what-is-esg. It is important to note that most ESG funds also use exclusions, or limits on fund-level

emissions. These effects on share prices may be greater than simple ESG scores.

3. Ronald Cohen has written a detailed take on the history and future of impact investing: *Impact: Reshaping Capitalism to Drive Real Change* (London: Ebury Press, 2020).

4. This is data from MSCI, one of the world's largest producers of aggregate measures of stock market performance, known as "indices". The link below is to an article looking at the performance during March 2020, when markets fell sharply due to the pandemic. It also includes five-year performance. Over five years ESG indices experienced better returns than the overall market, and fell by less than the overall market in March 2020. It is important to be very clear that although this narrative is often used by some to justify ESG investing, and for this reason it may have contributed to the wave in favour of ESG, there is no serious statistical information in these observations. They are based on small samples of data, and it is virtually impossible to identify the underlying causes. See "MSCI ESG indexes during the coronavirus crisis", 22 April 2020, https://www.msci.com/www/blog-posts/msci-esg-indexes-during-the/01781235361.

5. The new rules, which came into force in March 2021 – the EU's Sustainable Finance Disclosure Regulation (SFDR) – require all financial institutions who manage client savings and investments to disclose the ESG risks in these funds. The global equivalent, which is mandatory in Europe, but has gathered momentum through voluntary signatories is the Taskforce on Climate-Related Financial Disclosure (TCFD); see "What is the impact of the EU Sustainable Finance Disclosure Regulation (SFDR)?" S&P Global, 1 April 2021, https://www.spglobal.com/marketintelligence/en/news-insights/blog/what-is-the-impact-of-the-eu-sustainable-finance-disclosure-regulation-sfdr.

6. As of September 2021, Bloomberg data is based on 12-month forward consensus earnings.

7. BP PE ratio, Ycharts, https://ycharts.com/companies/BP/pe_ratio.

8. For some reason it is market convention to express equity valuations as a PE ratio; it is much more intuitive to invert this ratio and express it as an earnings yield. For example, a price/earnings multiple of 10x is also an earnings yield of 10 per cent. This is more intuitive because it is analogous to an interest rate. If a stock has an earnings yield of 10 per cent, effectively shareholders are earning 10 per cent per annum owning this stock, assuming the earnings are not falling. Using earnings yields as a metric allows one to easily compare the return on stocks to other assets – such as cash in the bank, bonds, or property.

9. These observations are based on market pricing in the first half of 2021.
10. Typically it is only very early stage, high-risk growth companies that finance investment through selling shares in their businesses. It is a common misconception that this implies that share prices don't affect investment rates. If a company trades on a high PE multiple because the market expects it to grow rapidly – for example Amazon – it is incentivized to maintain growth through investing in order to maintain its elevated PE multiple. So the PE multiple is affecting the rate of investment even though shares are not being used to directly finance investment.
11. Löffler, Petreski & Stephan, "Drivers of green bond issuance . . ." estimate that green bonds pay a lower interest rate on average of 0.4 per cent.
12. They do here. They can and do appear without hyphens elsewhere!
13. See "Rio Tinto exits coal with $2.25 billion Kestrel sale", Reuters, 27 March 2018, https://www.reuters.com/article/us-rio-tinto-plc-coal-divestiture-idUSKBN1H31WI. In a number of cases, those coal assets were sold to investment firms, some of which were run by former employees of Rio Tinto itself. According to Reuters, Rio Tinto sold its Kestrel coal mine to EMR Capital and Indonesia's Adaro Energy. EMR's board includes former Rio Tinto employees, see https://www.emrcapital.com/en-us/team.
14. Other global mining companies have acted in a similar way, with similar effects; see "Buy a coal mine, get a bonus: Glencore makes fortune on Colombian deal", Reuters, 27 October 2021, https://www.reuters.com/business/energy/buy-coal-mine-get-bonus-glencore-makes-fortune-colombian-deal-2021-10-27/.
15. J. Blas and J. Farchy, *The World For Sale: Money, Power, and the Traders Who Barter the Earth's Resources* (Oxford: Oxford University Press, 2021).
16. Blas & Farchy, *The World For Sale*, 17.
17. Extended producer responsibility is a good example of a smart regulation. It has primarily been deployed in electronics, automotives and consumer goods.
18. If you are a soft drinks manufacturer and you are liable for what happens to each plastic bottle at the end of its life you could be sued if it ends up in the wrong place. Not surprisingly, you think in a completely different way about how you design it, what you design, what materials you use and how you engage downstream in the product.
19. D. Rhys, "The surprisingly stunning afterlives of old coal plants", Bloomberg, 5 August 2021, https://www.bloomberg.com/news/features/2021-08-05/former-coal-plants-are-now-serving-lobster-growing-marijuana?sref=ZeZFqNJ3.

20. A. Currie, "Coal bad bank fund dangles complex climate kudos." Reuters, 5 August 2021, https://www.reuters.com/breakingviews/coal-bad-bank-fund-dangles-complex-climate-kudos-2021-08-05/.
21. For three strategies for financing decommissioning and examples, see M. Bazilian, B. Handler and K. Auth, "How public funders can help take coal offline ahead of schedule". Energy For Growth, 20 September 2021, https://www.energyforgrowth.org/memo/how-public-funders-can-help-take-coal-offline-ahead-of-schedule-three-financial-options/. For a broader examination of the decommissioning of energy assets, see D. Invernizzi *et al.* "Developing policies for the end-of-life of energy infrastructure: coming to terms with the challenges of decommissioning", *Energy Policy* 144 (2020), https://doi.org/10.1016/j.enpol.2020.111677 and P. Bodnar *et al.* (2020) "How to retire early: making accelerated coal phaseout feasible and just", Rocky Mountain Institute (2020), https://rmi.org/insight/how-to-retire-early.
22. In principle, we could estimate the value of the externality accumulated by Norway's oil sales, and put a specific number on the "reparations" that the Norges Fund should spend on decommissioning fossil fuels.
23. "Poland wins race to issue first green sovereign bond", Climate Bonds Initiative, 15 December 2016, https://www.climatebonds.net/2016/12/poland-wins-race-issue-first-green-sovereign-bond-new-era-polish-climate-policy.
24. Ranked 27th (of 28) on environmental policy by Sustainable Governance Indicators; see SGI, "Environmental policies" (2020), https://www.sgi-network.org/2020/Sustainable_Policies/Environmental_Policies.
25. "DWS shares slide after greenwashing claims prompt BaFin investigation", *Financial Times*, 26 August 2021, https://www.ft.com/content/0eb64160-9e41-44b6-8550-742a6a4b1022.
26. Bubbles in the stock market are less troublesome than bubbles in the banking sector, which tend to be good for bankers and bad for society.

4. SUPERCHARGE THE INDIVIDUAL

1. He puts it bluntly: "The next time someone tells you we must change behavior to reduce GHG [greenhouse gas] emissions, ask them how they changed behaviour to reduce emissions that were causing acid rain, smog, dispersion of lead, and destruction of the ozone layer. You will get a blank stare. No one changed behavior. Instead we changed technologies, with considerable success. We did this with

compulsory policies, especially regulations . . ." (Jaccard, *Citizen's Guide to Climate Success*, chapter 3).

2. C. Figueres and T. Rivett-Carnac, *The Future We Choose: The Stubborn Optimist's Guide to the Climate Crisis* (London: Manilla Press, 2021).

3. The Vegan Society, "Statistics", https://www.vegansociety.com/news/media/statistics.

4. For example, there is a lot of recent excitement about new payment technologies, such as cryptocurrencies. The irony is that the boring old credit card, which is a technology from the 1950s, is currently a more economically significant growth industry in economies like Germany. For reasons that aren't well understood, Germans are stubbornly wedded to the use of physical cash, despite a more efficient payment technology existing for decades. Similarly, Americans still rely heavily on paper cheques, a less efficient and reliable form of payment than electronic transfers. People don't simply adopt new technology because it exists.

5. Climate Watch, "Historical GHG emissions", https://www.climatewatchdata.org/ghg-emissions?end_year=2018&start_year=1990.

6. 2018 figures; see, H. Ritchie, "Climate change and flying", Our World in Data, 22 October 2020, https://ourworldindata.org/co2-emissions-from-aviation.

7. See E. Terrenoire *et al.*, "The contribution of carbon dioxide emissions from the aviation sector to future climate change", *Environmental Research Letters* 14:8 (2019), https://doi.org/10.1088/1748-9326/ab3086 and Climate Action Tracker, "Temperatures", https://climateactiontracker.org/global/temperatures/.

8. In 2019, annual worldwide plastic production created approximately one billion tonnes of CO_2e, equivalent to emissions from nearly 200 0.5GW coal plants, or to *c.*2 per cent of global GHG emissions; see CIEL, "Plastic and climate: the hidden costs of a plastic planet" (2019), http://www.ciel.org/wp-content/uploads/2019/05/Plastic-and-Climate-Executive-Summary-2019.pdf. This could increase by 250 per cent by 2040 if left unchecked, accounting for a fifth of the annual carbon budget. To limit warming to 1.5°C, see S. Reddy and W. Lau, "Breaking the plastic wave", PEW, 23 July 2020, https://www.pewtrusts.org/en/research-and-analysis/articles/2020/07/23/breaking-the-plastic-wave-top-findings.

9. For example, use of electric furnaces for monomer production, and contamination in waste flows reducing the quality of the recycled plastic, resulting in "downcycling".

10. C. Joyce, "Plastic has a big carbon footprint – but that isn't the whole story", NPR, 9 July 2019, https://www.npr.org/2019/07/09/

735848489/plastic-has-a-big-carbon-footprint-but-that-isnt-the-whole-story?t=1634552213661.

11. The presentation "How to stop the climate crisis in six months" (2021) by Extinction Rebellion co-founder, Roger Hallam, reveals a very rigorous perspective on how social change occurs: https://www.youtube.com/watch?v=CkZT0tSdUog. Similar arguments have also clearly informed Greta Thunberg's perspective; see her interview on the BBC's *Andrew Marr Show*, 31 October 2021, https://www.bbc.co.uk/iplayer/episode/m00116q5/the-andrew-marr-show-31102021.

12. Social norm theory is an example of this. Its origins lie in a 1986 study of alcohol consumption by university students (H. Perkins and A. Berkowitz, "Perceiving the community norms of alcohol use among students", *International Journal of the Addictions* 21:9/10, 961–76). The theory postulates that peer behaviours and beliefs are a significant driving factor for an individual's behaviour. Furthermore, the divergence of perceived versus actual norms of peer behaviour causes individuals to increase or decrease certain behaviours. The social norms approach addresses this misalignment between the perceived and the actual in order to organically shift behavioural patterns; see R. Dempsey, J. McAlaney and B. Bewick, "A critical appraisal of the social norms approach as an interventional strategy for health-related behavior and attitude change", *Frontiers in Psychology* 9, https://doi.org/10.3389/fpsyg.2018.02180 and P. Yamin *et al.*, "Using social norms to change behavior and increase sustainability in the real world", *Sustainability* 11:20 (2019), 5847, https://doi.org/10.3390/su11205847.

13. E. Chenoweth and M. Stephan, *Why Civil Resistance Works: The Strategic Logic of Nonviolent Conflict* (New York: Columbia University Press, 2011).

14. As Alison Taylor of BSR, a sustainability consultancy, observes: ". . . employees have embraced a whistleblowing model in which disclosing concerns to the public seems far more effective than a call to the internal ethics hotline" ("Exploring employee activism", Business for Social Responsibility blog, 27 May 2019, https://www.bsr.org/en/our-insights/blog-view/exploring-employee-activism-stakeholder-engagement).

15. "The more specific we can make demands for criminal justice and police reform, the harder it will be for elected officials to just offer lip service to the cause and then fall back into business as usual once protests have gone away", Barack Obama, Tweet, 1 June 2020, https://twitter.com/barackobama/status/1267534340503846912?lang=en.

16. There is a documented increase in employee activism across topics, see Herbert Smith Freehills, "Future of work: adapting to the

democratised workplace", https://www.herbertsmithfreehills.com/file/40326/download?token=_p7Oi8sL.

17. The Project Drawdown manual emphasizes finding groups of like-minded individuals for disproportionate impact, as well as mapping of power structures; see "Climate solutions at work", https://drawdown.org/publications/climate-solutions-at-work.

18. McKinsey & Co., "Overcoming obesity: an initial economic analysis", November 2014, https://www.mckinsey.com/~/media/mckinsey/business%20functions/economic%20studies%20temp/our%20insights/how%20the%20world%20could%20better%20fight%20obesity/mgi_overcoming_obesity_full_report.ashx.

19. Taxing sugary drinks has a much stronger moral and economic case; if we set a high enough price to discourage its consumption, individuals and society win, because its nutritional profile is empty. The case for taxing meat and dairy is more complicated in this respect.

20. Zeynep Tufekci presents a realistic perspective of the strengths and weaknesses of social media-based activism in *Twitter and Tear Gas: The Power and Fragility of Networked Protest* (New Haven, CT: Yale University Press, 2017).

21. See "Timberland's CEO on deciding to engage with angry activists", *Harvard Business Review*, 22 October 2013, https://hbr.org/2013/10/timberlands-ceo-on-deciding-to-engage-with-angry-activists.

22. See New Normal, "Friends of the Earth v Royal Dutch Shell – what did the Dutch Court rule, and what does it mean for Shell's business?", The pH Report, https://new-normal.com/wp-content/uploads/2021/06/Shell-v-Dutch-Friends-of-the-Earth-June-2021.pdf.

23. Shell's own target was to reduce the intensity of its greenhouse gas emissions by 20 per cent from 2016 levels.

24. "Shell to appeal landmark Dutch Court ruling on climate goals", Bloomberg Green, 20 July 2021, https://www.bloomberg.com/news/articles/2021-07-20/shell-to-appeal-landmark-climate-case-in-the-netherlands.

25. S. Subramanian, "Engine No. 1: the little hedge fund that shook Big Oil", Quartz, 28 May 2021, https://qz.com/2014413/engine-no-1-the-little-hedge-fund-that-shook-exxonmobil/ and H. Welsh, "Exxon's shareholder revolt is a warning for boards everywhere", Barron's, 21 May 2021, https://www.barrons.com/articles/exxons-shareholder-revolt-is-a-warning-shot-for-boards-everywhere-51622143164.

26. As of September 2021; see Exxon Mobil Corporation (XOM: NYSE) Stock Price & News (2021), Google Finance.

27. D. Centola, *Change: How to Make Big Things Happen* (London: John Murray, 2021).

28. When it comes to regulating food, research shows that the population expects the government to give guidance. We shouldn't underestimate the extent to which corporations and individuals can see regulation as solving problems for them, not interfering; see L. Wellesley and A. Froggatt, "Changing climate, changing diets: pathways to lower meat consumption", Chatham House, 24 November 2015, https://www.chathamhouse.org/2015/11/changing-climate-changing-diets-pathways-lower-meat-consumption.

5. SUPERCHARGE THE NATION

1. The vested interests of the US fossil fuel industry are a major obstacle to progress. For details of US coal's effective veto on legislation, see https://www.theguardian.com/commentisfree/2021/oct/29/joe-biden-climate-plan-emissions-cop26.
2. The ETC have stated that a 2030 target of 40 per cent electricity generation from wind and solar is both feasible and necessary to meet a mid-century goal of net zero, and estimate it is possible to run power systems with variable renewables at a max of 75–90 per cent. The biggest challenge is daily/seasonal balancing of variable systems which will rely in some part on existing fossil capacities. The ETC estimates roll-out of electrified economy (meeting 87 per cent of energy demands by direct and indirect electrification by 2050) could amount to $2.5 trillion per year, roughly 1.5 per cent of global GDP over the next 30 years; see ETC, "Making clean electrification possible". We believe widespread adherence to the policies we advocate would result in more ambitious results.
3. Typically, by low interest rates we mean countries with rates of less than 3 per cent. You can think of an interest rate as financial constraint; a low interest rate is a low level of constraint on capital expenditure. We include China in this category. See "World interest rates", Trading Economics, https://tradingeconomics.com/country-list/interest-rate?continent=world and "CO_2 emissions by country", World Population Review, https://worldpopulationreview.com/country-rankings/co2-emissions-by-country.
4. See "Interest rates: 2021 data; 2022 forecast; 1971–2020 historical", Trading Economics, https://tradingeconomics.com/united-kingdom/interest-rate. For Germany, Japan, France, Italy, see OECD, "Sovereign borrowing outlook for OECD countries 2020", https://www.oecd.org/finance/Sovereign-Borrowing-Outlook-in-OECD-Countries-2020.pdf. Even after the increases in bond yields in 2021, most developed economies can borrow at negative real interest rates for at least 20 years, using 2 per cent expected

inflation. For an explanation of real interest rates, see Federal Reserve Bank of San Francisco, "What it the difference between the real interest rate and the nominal interest rate?", https://www.frbsf. org/education/publications/doctor-econ/2003/august/real-nominal-interest-rate/.

5. The private sector in Europe is investing at required equity rates of return of 5–6 per cent, this means they have to make a 5–6 per cent return for investors, see IEA, *World Energy Investment 2020*, 167, https://iea.blob.core.windows.net/assets/ef8ffa01-9958-49f5-9b3b-7842e30f6177/WEI2020.pdf.

6. The obvious method is through guaranteeing producers minimum prices.

7. See M. Mazzucato, *The Entrepreneurial State: Debunking Public vs. Private Sector Myths* (London: Anthem Press, 2013).

8. Quantitative easing (QE) involves the central bank printing money and buying securities, such as government bonds, company bonds and equities. The idea is to bring down the yield, or interest rate, on these securities to provide a stimulus to the economy. QE should also be greened; we shall cover this in later notes.

9. A. Tooze, *Crashed: How a Decade of Financial Crises Changed the World* (London: Penguin, 2019) and *Shutdown: How Covid Shook the World's Economy* (London: Viking, 2021).

10. Inflation, the rate of change of prices throughout the economy, can vary significantly from one year to the next, particularly when the economy is adjusting to shocks like the financial crisis or the Covid-19 pandemic. For most of the past 20 years, rates of inflation have been below central bank targets, which are usually around 2 per cent; see: "Inflation, consumer prices (annual %) – Japan, European Union, United States (1960–2020)" [Dataset], World Bank, https://data.worldbank.org/indicator/FP.CPI.TOTL. ZG?locations=JP-EU-US.

11. Given anxieties about inflation, it is important to remember that for most of the past 20 years, central banks have typically failed to generate sufficient inflation to meet their targets. While it's possible that this has changed post-Covid, it is prudent to have contingency plans to deal with a persistent risk of inflation being too low, which has been dominant for a long time. At the time of writing (October 2021), Covid-related supply-chain disruption and the shift in spending in favour of goods at the expense of services during lockdown has caused a spike in measured price inflation, this is unlikely to persist for more than a year-or-so, as the global economic system adapts. One of the clearest and most insightful analyses is provided by Matt Klein, author of "The Overshoot", see https:// theovershoot.co/p/paying-the-covid-bill.

12. Many central banks provide loans to the commercial banking system with strings attached, as we describe, although not targeted specifically at sustainable energy. Some of these also lend at negative interest rates. A majority of the developed world's largest central banks launched targeted lending schemes during the pandemic. No one has yet combined the best of all approaches. The Bank of Japan, is the only major central bank we are aware of which has repurposed its term funding scheme to finance sustainable investments in the energy sector. But Japan has not gone far enough with negative interest rates to really move the dial. Its loans are priced at zero. To our knowledge, the ECB used the most aggressive reduction in interest rates, lending to banks at rates as low as −1 per cent. But in Europe, the conditions for access have been limited to aggregate lending and not targeted at the sustainable energy sector, like in Japan. The Bank of Israel made its lending conditional on the preferential rates being passed on to customers. A "best-of-all" programme would be at sharply negative interest rates, say −2 or −3 per cent, targeted at sustainable investment, and banks would be required to pass on something like 50 per cent of the reduced interest rates to the final borrower.

13. See L. Kihara and D. Leussink, "BOJ rolls out climate scheme, to disburse first loans late December". Reuters, 22 September 2021, https://www.reuters.com/business/sustainable-business/boj-says-start-accepting-bank-applications-climate-scheme-2021-09-22/ and G. Caswell, "Bank of Japan to launch climate lending facility", Green Central Banking, 21 June 2021, https://greencentralbanking.com/2021/06/21/bank-of-japan-to-launch-climate-lending-facility/.

14. "Inflation, consumer prices (annual %) – Japan (1960–2020)". [Dataset] World Bank, https://data.worldbank.org/indicator/FP.CPI.TOTL.ZG?locations=JP.

15. Recessions have tended to strike developed economies at least once a decade over the past 50 years.

16. J. Williams, "Measuring the effects of monetary policy on house prices and the economy", BIS Papers 88 (2016), https://www.bis.org/publ/bppdf/bispap88_keynote.pdf.

17. The evidence so far from these programmes is that there is no free-riding. It is clear that conditions need to be attached, the data should be made publicly available, and the central banks need oversight. It is also important to be clear that central banks are really not taking on much risk at all under these programmes, they are simply reducing the cost of capital to the sector and providing an economic stimulus. The ECB has a lot of research with evidence suggesting a material improvement in financial conditions in response to TLTRO; see

D. Andreeva and M. García-Posada, "The impact of the ECB's targeted long-term refinancing operations on banks' lending policies", ECB Working Paper 2364 (January 2020), https://www.ecb. europa.eu/pub/pdf/scpwps/ecb.wp2364–12a4540091.en.pdf. The Bank of Israel has also specifically tied the loans to a pass-through to interest rates to the underlying borrower; see "Bank of Israel launches targeted lending operations", Central Banking, 6 April 2020, https://www.centralbanking.com/central-banks/ monetary-policy/monetary-policy-decisions/7521211/bank-of-israel-launches-targeted-lending-operations. Clearly, the regulatory clout of the central bank has to be effective to manage this risk.

18. Regarding the risk central banks are taking with these targeted lending programmes it is important to remember that these facilities were originally introduced to overcome a technical problem that economists have been concerned about, which is the effective lower bound of interest rates. In terms of the financial risk central banks are taking, we need to remember that central banks are lending to commercial banks who then take all the credit risk of the energy companies and the projects which are being financed. There is very, very little financial risk to the central bank, and in many instances, these loans are also protected by collateral. See also, E. Lonergan and M. Greene, "Dual interest rates give central banks limitless fire power", VoxEU, 3 September 2020, https://voxeu.org/article/dual-interest-rates-give-central-banks-limitless-fire-power.

19. When a bank makes a loan it is required by law to put aside capital. This is a way of protecting depositors. Banks take deposits and make loans (strictly-speaking, by making loans they create deposits). If their loans go bad, they don't have the resources to honour deposits. To ensure that deposits are safer, the regulator requires the banks to hold more capital against riskier loans.

20. Some central bankers will have reservations about using capital ratios to direct lending towards different sectors. The purpose of capital ratios is to ensure banks don't take on too much risk, and that systemic risk is mitigated. A couple of observations are pertinent. The risks associated with lending to fossil fuel industries may in fact be greater than standard financial metrics suggest. These are sectors in decline, and policies such as those outlined in this book do not bode well for the financial prospects of carbon emitters. Secondly, there are systemic considerations which bank regulators should really care about. The economic consequences of pronounced climate change could be disastrous for the banking sector, walking blindfolded into a fire is hardly sound regulation.

21. For more detail, see E. Lonergan, "QE is debt reduction", https://www.philosophyofmoney.net/qe-is-debt-reduction/.

22. B. Wink, "The world's major central banks bought $1.4 trillion of assets in March – 5 times the last record set after the financial crisis", Markets Insider, 21 April 2020, https://markets.businessinsider.com/news/stocks/central-banks-buy-trillion-financial-assets-g7-march-federal-reserve-2020-4.
23. For more on the just transition, see "What is a just transition?", European Bank for Reconstruction and Development, https://www.ebrd.com/what-we-do/just-transition.
24. Many sovereign wealth funds have poor governance and do not publish returns. However, the opposite is also the case. It is very clear how a national endowment which adheres would operate. Where returns are made public, they have all exceeded the target return we recommend, see A. Bauer, "How good are sovereign wealth funds at investing money made from natural resources?", Natural Resource Governance Institute, 13 June 2018, https://resourcegovernance.org/blog/how-good-are-sovereign-wealth-funds-investing-money-made-natural-resources. Strictly speaking, in considering investment returns, one should focus on real interest rates. Real interest rates is the interest rate we pay minus the rate of inflation. We can compare this to the real return on our investments.
25. For details on fund performance, see https://www.nbim.no/en/the-fund/market-value/ (Norges Fund), https://wellcome.org/press-release/wellcome-releases-201920-annual-results (Wellcome Trust) and https://www.hmc.harvard.edu/partners-performance/ (Harvard's Endowment).
26. See E. Lonergan, "Safe asset issuance and the discovery of oil", https://www.philosophyofmoney.net/safe-asset-issuance-discovery-oil/.
27. Not all sovereign wealth funds publish their returns. This sample list indicates average annual nominal return between 2010 and 2017 of between 3 and 15.5 per cent; https://resourcegovernance.org/blog/how-good-are-sovereign-wealth-funds-investing-money-made-natural-resources.
28. For example, President Biden has announced plans for US electricity generation to reach net zero by 2035. See "Path to net-zero: US utilities in no rush to meet Biden's 2035 climate goal", S&P Global, 16 June 2021, https://www.spglobal.com/marketintelligence/en/news-insights/latest-news-headlines/path-to-net-zero-us-utilities-in-no-rush-to-meet-biden-s-2035-climate-goal-64891626.
29. See T. Conlan and P. Regan, "Implementing the 2020 stimulus: lessons from the 2009 Recovery Act", Government Executive, 13 April 2021, https://www.govexec.com/oversight/2020/04/implementing-2020-stimulus-lessons-2009-recovery-act/

164319/; D. Carpenter and D. Moss, *Preventing Regulatory Capture: Special Interest Influence and How to Limit it* (Cambridge: Cambridge University Press, 2013); OECD, *Preventing Corruption in Public Procurement* (2016), https://www.oecd.org/gov/ethics/Corruption-Public-Procurement-Brochure.pdf and F. Rojas, "Recovery Act transparency", IBM Center for the Business of Government (2012), https://www.businessofgovernment.org/sites/default/files/Recovery%20Act%20Transparency.pdf.

30. This is only considering stranded assets in a transition from a hydrocarbon based economy to a decarbonized one. There may well be other stranded assets in a wholesale green transition, relating to a number of industries, including plastics, textiles and chemicals. In terms of fossil fuel related stranded assets, 300 publicly listed companies own 98 per cent of fossil fuel reserves and production. These 300 represent *c*.$5 trillion in market cap. Total global stock of wealth is approximately $435 trillion. See "Who owns the world's fossil fuels?", Influence Map, https://influencemap.org/FossilFuel300 and J.-F. Mercure *et al.*, "Macroeconomic impact of stranded fossil fuel assets", *Nature Climate Change* 8:7, 588–93, https://doi.org/10.1038/s41558-018-0182-1.

31. For data on employment in the energy sector, see Czako, "Employment in the energy sector". There is an extensive literature on the "just transition" to ensure our climate response delivers inclusive growth, from legal, ethical, climate science and policy perspectives. Simone Abram *et al.* (2020) provide a summary of success factors for a just climate transition. They emphasise that transition needs to be *perceived* as just to avoid social backlash against decarbonization, and recommend deliberative, inclusive policy processes to achieve that. See S. Abram *et al.*, "Just transition: pathways to socially inclusive decarbonisation", COP26 Universities Network Briefing, October 2020, https://www.ucl.ac.uk/public-policy/just-transition-pathways-socially-inclusive-decarbonisation. The EU has led the way in structuring financing mechanisms and allocating meaningful budget towards supporting social justice as part of this transition, through the just transition mechanism (JTM). For an overview, see "The just transition mechanism: making sure no one is left behind", European Commission, https://ec.europa.eu/info/strategy/priorities-2019-2024/european-green-deal/finance-and-green-deal/just-transition-mechanism_en.

32. We also need to eliminate subsidies for fossil fuels. Global fossil fuel subsidies were $370 billion in 2019 vs $100 billion for renewables; see J. Casey and Y. Cholteeva, "Debate: subsidies versus carbon pricing in European renewables", Power Technology,

8 April 2021, https://www.power-technology.com/features/
debate-subsidies-versus-carbon-pricing-in-european-renewables/.

33. C. Perez, *Technological Revolutions and Financial Capital: The Dynamics of Bubbles and Golden Ages* (Cheltenham: Elgar, 2003).

34. On sensible ways to raise tax, see "Don't waste a good crisis: reform taxes to make tax rises less painful", Institute of Fiscal Studies, https://ifs.org.uk/publications/14989.

35. Incentives are aligned to some degree, but there is misalignment in rental properties, where landlords, who would front capital costs, do not see energy bill savings.

36. Such as Canada who included almost CA$1 billion in the 2019 budget for municipalities to increase energy efficiency in buildings, and New Zealand who committed $500 million for building retrofits, see UNEP, *2020 Global Status Report for Buildings and Construction*, https://globalabc.org/sites/default/files/inline-files/2020%20 Buildings%20GSR_FULL%20REPORT.pdf. See also Department of Energy & Climate Change, "Green Deal Finance Company funding to end", press release, 23 July 2015, https://www.gov.uk/government/ news/green-deal-finance-company-funding-to-end.

37. The effect of EPICs in solar take up is very clear. Unfortunately, governments have often reacted to high success rates by terminating incentives.

38. F. Harvey, "Buyers of brand-new homes face £20,000 bill to make them greener", *The Guardian*, 25 August 2021, https://www. theguardian.com/environment/2021/jan/23/buyers-of-brand- new-homes-face-20000-bill-to-make-them-greener.

39. At an aggregate level, building sector energy intensity has been decreasing 0.5–1 per cent/yr. Despite this, aggregate CO_2 emissions have been growing since 2010. Drivers of this CO_2 growth are increasing population – i.e. more buildings; increasing energy demand due to extreme weather around the world; and increasing floor area of buildings (growing at 2.5 per cent per year, so outcompeting energy intensity improvements by 1.5–2 per cent). See IEA, "Tracking Buildings 2020", https://www.iea.org/reports/ tracking-buildings-2020.

40. There is little consensus among experts. The ETC thinks more GHG reduction potential in next decade in energy efficiency of buildings (1.6–2.1 Gt/yr) versus electrification (1 Gt/yr). IRENA very much agrees with most buildings subsidies to renewables; see M. Taylor, *Energy Subsidies: Evolution in the Global Energy Transformation to 2050* (Abu Dhabi: IRENA, 2020), www.irena.org/-/media/Files/IRENA/ Agency/Publication/2020/Apr/IRENA_Energy_subsidies_2020.pdf. Our best guess is that IRENA is probably right, but both should be pursued.

41. For example, Mexico and the Netherlands; see IEA, *Green Mortgages – Policies* (2012), https://www.iea.org/policies/3884-green-mortgages and *Green Mortgage Programme – Policies* (2017), https://www.iea.org/policies/1168-green-mortgage-programme?country=Mexico%2CNetherlands&page=1®ion=North%20America&status=In%20force&topic=Energy%20Efficiency&type=Loans%20%2F%20debt%20finance.

42. A. Tooze, "Green mortgages: homes need to catch up to climate change", The Hill, 13 May 2021 https://thehill.com/opinion/finance/553175-green-mortgages-homes-need-to-catch-up-to-climate-change.

43. So far, many of the models we have seen link the mortgage to measures of energy efficiency, when we really want them to be linked to emissions over time. So typically, the emphasis is on some form of insulation, which often involves substantial disruption, which means it won't happen. New builds are clearly most tractable and smart regulations are key here – energy efficiency, electrification and future proofing needs to be regulated for new builds, and green mortgages made available. But the real challenge is collapsing the emissions of the existing housing stock. Here there needs to be a converted focus on remortgaging at preferential rates, and capturing sales. Minimizing disruption and inconvenience is mission-critical and is often at odds with energy efficiency. There is a natural time lag between purchasing a property and moving in. So there is an opportunity to make an investment in retrofitting and adding insulation, and doing this without the disruption of living at the new property. We are introducing EPICs both from the interest rate cost on the mortgage but also on the tax exemption. This could see 80–90 per cent of all new mortgages resulting in 20 per cent energy efficiency across the residential housing sector. Furthermore, we could also see programmes of refinancing which could also come with common incentives. The emphasis on emissions reduction, electrification and utilizing the rapid expansion of technological options is critical here. The residential sector accounts for around 60 per cent of the emissions of buildings, but mortgages, smart regs and negative taxes are also the key to transforming commercial property.

44. OECD, "Tax on property (2000–2020)" [Dataset], https://data.oecd.org/tax/tax-on-property.htm.

45. From July 2020 to June 2021, the UK government introduced a "stamp duty holiday", whereby the value of any properties under £500,000 was exempt from the standard rates, which are a progressive tax on the value of a property above the first £125,000.

46. For example, CAFE standards in the US; see Y. Wang and Q. Miao,

"The impact of the corporate average fuel economy standards on technological changes in automobile fuel efficiency", *Resource and Energy Economics* 63 (2021), 101211, https://doi.org/10.1016/j.reseneeco.2020.101211.

47. "Domestic private rented property: minimum energy efficiency standard – landlord guidance", GOV.UK, 4 May 2020, https://www.gov.uk/guidance/domestic-private-rented-property-minimum-energy-efficiency-standard-landlord-guidance#enforcement-and-penalties.
48. See Gates, *How to Avoid a Climate Disaster*, 192–4.
49. Approximately 26 million jobs/yr in variable renewables by 2050. See ETC, "Making clean electrification possible", 64.

6.   SUPERCHARGE THE WORLD

1. M. Blyth, "My only line on the election. The era of neoliberalism is over. The era of neo-nationalism has just begun." Tweet, 9 November 2016, https://twitter.com/mkblyth/status/796392986011639810?lang=en.
2. The total number of international migrants at mid-year 2020 was 280.6 million, up from 173 million at mid-year 2000; see Migration Data Portal, https://www.migrationdataportal.org/international-data?i=stock_abs_&t=2020. In terms of global trade, import and export value of goods in 2019 increased to $18.5 trillion and $20.5 trillion respectively, up from $16.1 trillion and $17.7 trillion in 2008; see "World exports, imports, tariff by year", World Bank, https://wits.worldbank.org/CountryProfile/en/Country/WLD/Year/LTST/Summary.
3. Helm, *Net Zero*.
4. A. Lieven, *Climate Change and the Nation State: The Case for Nationalism in a Warming World* (Oxford: Oxford University Press, 2020).
5. All of these countries are insufficient or highly insufficient on the Climate Action Tracker, see https://climateactiontracker.org.
6. Danish Ministry of Climate, Energy and Utilities, *Denmark's Integrated National Energy and Climate Plan* (December 2019), https://ec.europa.eu/energy/sites/ener/files/documents/dk_final_necp_main_en.pdf and *Denmark's Mid-century, Long-term Low Greenhouse Gas Emission Development Strategy* (December 2020), https://unfccc.int/sites/default/files/resource/ClimateProgramme2020-Denmarks-LTS-under-the%20ParisAgreement_December2020_.pdf.
7. The positive contribution to Denmark's employment and exporting capacity is notable; see "Employment, export and revenue", Wind Denmark, https://en.winddenmark.dk/wind-in-denmark/statistics/employment-export-and-revenue and "Denmark sets record by

sourcing nearly half its power from wind energy", *The Independent*, 3 January 2020, https://www.independent.co.uk/news/world/europe/denmark-power-wind-energy-climate-turbines-sea-a9269076.html.

8. J. Pyper, "How long will coal remain king in India?", Wood Mackenzie, 20 January 2021, https://www.greentechmedia.com/articles/read/coal-king-india.

9. See "Data Explorer: historical emissions" [Dataset], Climate Watch, https://www.climatewatchdata.org/data-explorer/historical-emissions?historical-emissions-data-sources=cait&historical-emissions-gases=all-ghg&historical-emissions-regions=All%20Selected&historical-emissions-sectors=total-including-lucf&page=1.

10. Installed renewable power generation capacity in India achieved a CAGR of 17.33 per cent between 2016 and 2020; India has set a target of 175GW installed capacity of renewable energy by 2022, see IBEF, *Renewable Energy Industry in India: Overview, Market Size & Growth* (August 2021), https://www.ibef.org/industry/renewable-energy.aspx.

11. India is the second largest importer of coal despite having the world's fourth largest reserves, and coal powers over 70 per cent of the country's electricity demand; see H. Ritchie, "India: CO_2 country profile". Our World in Data, 11 May 2020, https://ourworldindata.org/co2/country/india. On growth in India's coal consumption see "India's oil, coal addiction hurdle for speeding up emission goals", S&P Global Platts, 18 August 2021, https://www.spglobal.com/platts/en/market-insights/latest-news/energy-transition/081821-lower-emissions-a-concern-for-india-as-appetite-for-coal-oil-rises-with-energy-demand.

12. S. Gross, "Coal is king in India – and will likely remain so", Brookings, 8 March 2019, https://www.brookings.edu/blog/planetpolicy/2019/03/08/coal-is-king-in-india-and-will-likely-remain-so/.

13. India accounts for 18 per cent of the world's premature deaths related to air pollution, see IEA, "Premature deaths related to air pollution in India, 2015, 2019 and 2040 by scenario" (2021), https://www.iea.org/data-and-statistics/charts/premature-deaths-related-to-air-pollution-in-india-2015-2019-and-2040-by-scenario. Recent research indicates it is not just an urban problem; see A. Ravishankara *et al.*, "Outdoor air pollution in India is not only an urban problem", *Proceedings of the National Academy of Sciences* 117:46 (2020), 28640–44, https://doi.org/10.1073/pnas.2007236117

14. Government of India Ministry of Steel, "Energy and environment management in steel sector" (2021), https://steel.gov.in/energy-environment-management-steel-sector.

15. "Installed power from renewable sources by type Spain 2019" [Dataset], Statista, 2 July 2021, https://www.statista.com/statistics/1031674/installed-power-capacity-from-renewable-sources-in-spain/.

16. A review of stagnation in the growth of wind capacity identifies "the shift to auction route that seeks the lowest per-unit cost in the wind sector, lack of financial incentives and difficulties in finding land for the projects"; see M. Aggarwal, "India's struggling wind power sector needs fresh air to regain growth", Mongabay, 5 August 2021, https://india.mongabay.com/2021/08/indias-struggling-wind-power-sector-needs-fresh-air-to-regain-growth/.

17. The carbon intensity of India's power sector is 725gm of CO_2 per kilowatt-hour (gCO_2/kWh), compared with a global average of 510 gCO_2/kWh, underlining the predominant role of inefficient coal-fired generation; see IEA, *India Energy Outlook 2021*, https://iea.blob.core.windows.net/assets/1de6d91e-e23f-4e02-b1fb-51fdd6283b22/India_Energy_Outlook_2021.pdf.

18. Growing at a CAGR of 15 per cent; see IBEF, *Renewable Energy Industry in India*.

19. IEA, *India Energy Outlook 2021*.

20. G. Shrimali, "Financial performance of renewable and fossil power sources in India", *Sustainability* 13 (2021), https://doi.org/10.3390/su13052573.

21. As of 5 October 2021, the yield to maturity (effectively the fixed rate of interest) on Danish 20-year bonds was 0.23 per cent; see Market Watch, "TMBMKDK-20Y: Denmark 20-year government bond overview" [Dataset], 25 October 2021, https://www.marketwatch.com/investing/bond/tmbmkdk-20y?countrycode=bx.

22. The eurozone, including the UK, borrowed close to $2 trillion in 2020. For the total in the eurozone, see "Euro zone debt surges in 2020 on pandemic spending", Reuters, 22 April 2021, https://www.reuters.com/article/eurozone-debt-idUSL8N2MF2ZC and for the UK, M. Keep, "Government borrowing: peacetime record confirmed", House of Commons Library, 23 April 2021, https://commonslibrary.parliament.uk/government-borrowing-peacetime-record-confirmed/.

23. These ideas were originally co-developed with the brilliant Angus Armstrong, Director of the research network, Rebuilding Macroeconomics at UCL. We are entirely responsible for any errors. See Rebuilding Macroeconomics, https://www.ucl.ac.uk/pals/research/clinical-educational-and-health-psychology/research-groups/centre-decision-making-uncertainty/rebuilding-macroeconomics.

24. Vestas, Denmark's wind turbine manufacturer being a case in

point, in both India and Vietnam. The latter with financing from the Danish government, a precise example of the type of policy we are advocating, but on a much greater scale. See "Vestas bags 101MW India order", Renewable Energy News, 21 September 2021, https://renews.biz/72372/vestas-bags-101mw-india-order/, A Gupta, "Vestas and Danish Export Credit Agency unlock Vietnamese wind energy project through innovative financing structure that can be used in other emerging markets", *EQ International*, 13 March 2019, https://www.eqmagpro.com/vestas-and-danish-export-credit-agency-unlock-vietnamese-wind-energy-project-through-innovative-financing-structure-that-can-be-used-in-other-emerging-markets/.

25. Looking at the top five exporters of wind turbine exports to India, the Indian market dominates their export market. For share of exports to India vs rest of world: Denmark (96%); China (69%); Spain (75%); Germany (45%); Italy (98%), see "Wind turbine import trade database" [Dataset], Infodrive India, https://www.infodriveindia.com/shipment-data/india-import-data-of-wind+turbine.

26. Various meta-studies of economic crisis over the last 200 years indicate a consistent increase in polarization and support of extremes, particularly extreme right. See J. Cohen-Setton, "The predictability of political extremism". Bruegel, 14 December 2015, https://www.bruegel.org/2015/12/the-predictability-of-political-extremism/ and M. Funke, M. Schularick and C. Trebesch, "Going to extremes: politics after financial crises, 1870–2014". SSRN Electronic Journal. Published 2015. https://doi.org/10.2139/ssrn.2688897.

27. UNHCR estimates 20 million annual climate refugees, indicating just one lens on the humanitarian impacts. Estimates for the total number of climate refugees by 2050 extend into the billions; see UNHCR, "Climate change and disaster displacement", https://www.unhcr.org/uk/climate-change-and-disasters.html and "The climate crisis, migration, and refugees", Brookings, July 2019, https://www.brookings.edu/research/the-climate-crisis-migration-and-refugees/.

28. For more on the institutional function of money as the antithesis of nationalism, see E. Lonergan, *Money*, second edition (London: Routledge, 2014), written in the wake of the global financial crisis.

29. IMF, "Questions and answers on Special Drawing Rights", https://www.imf.org/en/About/FAQ/special-drawing-right.

30. The current split is US dollars (42%), euros (31%) and renminbi (11%), and the balance is split between Japanese yen and pounds sterling; see IMF, "Special Drawing Rights (SDR) factsheet", https://www.imf.org/en/About/Factsheets/Sheets/2016/08/01/14/51/Special-Drawing-Right-SDR.

31. IMF, "IMF governors approve a historic US$650 billion SDR allocation of Special Drawing Rights", press release, 2 August 2021, https://www.imf.org/en/News/Articles/2021/07/30/pr21235-imf-governors-approve-a-historic-us-650-billion-sdr-allocation-of-special-drawing-rights.

32. IMF, "The IMF and the World Bank" (2021), https://www.imf.org/en/About/Factsheets/Sheets/2016/07/27/15/31/IMF-World-Bank.

33. For the history of ECAs see IMF, "The changing role of export credit agencies", https://www.imf.org/external/pubs/nft/1999/change/.

34. J. Basquill, "Australia's export credit agency under pressure following rising support for fossil fuels", *Global Trade Review*, 7 July 2021, https://www.gtreview.com/news/sustainability/australias-export-credit-agency-under-pressure-following-rising-support-for-fossil-fuels/.

35. For a perceptive history of Saudi Arabia's stance on climate change, see L. Fang and S. Lerner, "Saudi Arabia denies its key role in climate change even as it prepares for the worst", The Intercept, 18 September 2019, https://theintercept.com/2019/09/18/saudi-arabia-aramco-oil-climate-change/. On Russia, see Y. Fedorinova, "Putin says Russia will target carbon neutrality by 2060", Bloomberg Green, 13 October 2021, https://www.bloomberg.com/news/articles/2021-10-13/putin-says-russia-will-target-carbon-neutrality-by-2060?sref=ZeZFqNJ3.

36. Russia accounts for *c*.4 per cent of worldwide emissions whilst Saudi Arabia emits 1.3 per cent of global GHGs, see Climate Watch historical emissions dataset.

37. "Leading steel importers globally by country 2019" [Dataset], Statista, https://www.statista.com/statistics/650538/leading-steel-importers-globally-sorted-by-country/.

38. For a superb analysis of the Chinese government's policies around emissions reductions and the coal industry, see M. Klein's "Appeasing the Chinese government 'for the climate' makes no sense", The Overshoot, 22 October 2021, https://theovershoot.co/p/appeasing-the-chinese-government.

39. H. Ritchie, "Who has contributed most to global CO_2 emissions?", Our World in Data, 1 October 2019, https://ourworldindata.org/contributed-most-global-co2.

40. Klein, "Appeasing the Chinese Government".

41. *Ibid*.

42. S. Mallapaty, "How China could be carbon neutral by mid-century", *Nature* 586 (2020), 482–3, https://doi.org/10.1038/d41586-020-02927-9; S. Ladislaw and N. Tsafos, "Beijing is winning the race to build – and sell – clean energy technology", *Foreign Policy*, 2 October

2020, https://foreignpolicy.com/2020/10/02/china-clean-energy-technology-winning-sell/.

43. On the speed with which China reached dominance of the global solar industry and the effect of its use of EPICs, see J. Fialka, "Why China is dominating the solar industry", *Scientific American*, 19 December 2016, https://www.scientificamerican.com/article/why-china-is-dominating-the-solar-industry/. For the role of coal in China's solar industry, see M Dalton, "Behind the rise of U.S. solar power, a mountain of Chinese coal", *Wall Street Journal*, 31 July 2021, https://www.wsj.com/articles/behind-the-rise-of-u-s-solar-power-a-mountain-of-chinese-coal-11627734770. On the evidence of human rights abuses, see L. Murphy and N. Elima, "In broad daylight: Uyghur forced labour and global solar supply chains" (May 2021), Sheffield Hallam University Helena Kennedy Centre for International Justice, https://www.shu.ac.uk/helena-kennedy-centre-international-justice/research-and-projects/all-projects/in-broad-daylight.

44. The 14th Five-Year Plan covers the period 2021–25; see "China cuts 'carbon intensity' 18.8% in past five years, in effort to rein in emissions", Reuters, 2 March 2021, https://www.reuters.com/article/us-china-climatechange/china-cuts-carbon-intensity-18-8-in-past-five-years-in-effort-to-rein-in-emissions-idUSKBN2AU157.

45. In 2021, China introduced an emissions trading, cap-and-trade system, covering 30 per cent of industrial emissions; see B. Nogrady, "China launches world's largest carbon market: but is it ambitious enough?", *Nature* 595, 637 (2021), https://doi.org/10.1038/d41586-021-01989-7. However, given how low the effective carbon price is, between $6 and $10 per ton, a far cry from the Stiglitz-Stern recommendation of $50 rising to $100, it's unlikely to shift emissions. Economists Joseph Stiglitz and Nicholas Stern co-chair the High-Level Commission on Carbon Prices, tasked with exploring carbon price levels required to change behaviour. Their 2017 report established a $50–100 price range; see *Report of the High-Level Commission on Carbon Prices* (2017), https://static1.squarespace.com/static/54ff9c5ce4b0a53decccfb4c/t/59b7f2409f8dce5316811916/1505227332748/CarbonPricing_FullReport.pdf.

46. See S. Tabeta, "China plans to phase out conventional gas-burning cars by 2035", Nikkei Asia, 27 October 2020, https://asia.nikkei.com/Business/Automobiles/China-plans-to-phase-out-conventional-gas-burning-cars-by-2035 and S. Kang, "China aims for EVs to account for 50% of all car sales by 2035", S&P Global Market Intelligence, 29 October 2020, https://www.spglobal.com/

marketintelligence/en/news-insights/latest-news-headlines/
china-aims-for-evs-to-account-for-50-of-all-car-sales-by-
2035-60954964.

47. IEA, "Country profile: China" (2021), https://www.iea.org/countries/
china.

48. "New report shows a high ambition coal power phaseout in China is
feasible", Center for Global Sustainability, University of Maryland,
6 January 2020, https://cgs.umd.edu/news/new-report-shows-
high-ambition-coal-power-phaseout-china-feasible.

49. Intriguingly, its attempts to reduce its reliance on coal may have
been a factor behind the spike in global gas prices in 2021.

50. An EU-centric addendum: for recommendations on how to generate
a constructive "triumvirate" of competition, rivalry and cooperation
in EU–China climate relations, see section on "Recommendations
for EU action" in J. Oertel, J. Tollmann and B. Tsang, "Climate
superpowers: how the EU and China can compete and cooperate for
a green future", ECFR, 3 December 2020, https://ecfr.eu/publication/
climate-superpowers-how-the-eu-and-china-can-compete-and-
cooperate-for-a-green-future/#recommendations-for-eu-action.

51. By 2017, 75 per cent of Belt and Road Initiative projects were power
and transport infrastructure. In 2021, renewables made up over
50 per cent of BRI energy investments, and has exited two planned
coal projects, in Bangladesh and Zimbabwe; see D. de Boer, C. Wang
and F. Danting, "Good progress in greening the BRI", CCICED, 20 July
2021, https://cciced.eco/climate-governance/good-progress-in-
greening-the-bri/.

7. WHY THIS IS NOT A BOOK ABOUT TREES

1. D. Carrington, "Tree planting 'has mind-blowing potential' to tackle
climate crisis", *The Guardian*, 31 August 2021, https://www.
theguardian.com/environment/2019/jul/04/planting-billions-
trees-best-tackle-climate-crisis-scientists-canopy-emissions.

2. Greta Thunberg: "Yes, of course we need to plant as many trees as
possible. Yes, of course we need to keep the existing trees standing
and rewild and restore nature. But there's absolutely no way around
stopping our emissions of greenhouse gases and leaving fossil fuels
in the ground", Twitter, 5 July 2019.

3. It is worth bearing in mind that this is significantly impacting the
pH of the ocean, effectively making it more acidic, but also that the
ocean is only absorbing CO_2 because we are producing so much.
Prior to industrialization, the ocean was actually a net source of CO_2.

4. N. Harris and D. Gibbs, "Forests absorb twice as much carbon as they
emit each year", World Resources Institute, 21 January 2021, https://

www.wri.org/insights/forests-absorb-twice-much-carbon-they-emit-each-year.

5. FAO, *State of the World's Forests 2020*, https://www.fao.org/state-of-forests/en/.
6. The rate of net deforestation declined from 7.8 million ha per year in the decade 1990–2000 to 4.7 million ha per year in 2010–2020; FAO, *Global Forest Resources Assessment 2020*, https://www.fao.org/forest-resources-assessment/2020/en/. The improvement in deforestation rates is largely attributable to REDD+ sponsored programmes.
7. For a detailed read on the complexity of the Amazon, and how to think about its climate impacts as a system, see C. Welch, "First study of all Amazon greenhouse gases suggests the damaged forest is now worsening climate change", *National Geographic*, 11 March 2021, https://www.nationalgeographic.com/environment/article/amazon-rainforest-now-appears-to-be-contributing-to-climate-change.
8. "Lula retains solid lead over Bolsonaro for 2022 Brazil race, poll shows", Reuters, 17 September 2021, https://www.reuters.com/world/americas/lula-retains-solid-lead-over-bolsonaro-2022-brazil-race-poll-shows-2021-09-17/.
9. Examples include the Climate Principals' Amazon Protection Plan, https://climateprincipals.org/amazon-plan/.
10. M. Evans, "Effective incentives for reforestation: lessons from Australia's carbon farming policies", *Current Opinion in Environmental Sustainability* 32 (2018), 38–45, https://doi.org/10.1016/j.cosust.2018.04.002 and "The Reforestation Accelerator: a powerful tool for driving natural climate solutions", The Nature Conservancy, 20 May 2020, https://www.nature.org/en-us/what-we-do/our-insights/perspectives/reforestation-accelerator-driving-natural-climate-solutions/.
11. Zhao, B. *et al.*, "North American boreal forests are a large carbon source due to wildfires from 1986 to 2016". *Scientific Reports* 11, 7723 (2021), https://doi.org/10.1038/s41598-021-87343-3.
12. McKinsey & Co., "Climate math: what a 1.5-degree pathway would take", McKinsey Quarterly. April 2020, https://www.mckinsey.com/~/media/McKinsey/Business%20Functions/Sustainability/Our%20Insights/Climate%20math%20What%20a%201%20point%205%20degree%20pathway%20would%20take/Climate-math-What-a-1-point-5-degree-pathway-would-take-final.pdf.
13. K. Harrington, "World's largest carbon capture project launches in Texas", AIChE, 31 January 2017, https://www.aiche.org/chenected/2017/01/worlds-largest-carbon-capture-project-launches-texas.

14. For example, StartUs Insights, "Top 5 carbon capture and storage start-ups impacting the energy sector", https://www.startus-insights.com/innovators-guide/5-top-carbon-capture-storage-startups-impacting-the-energy-sector/ and E. Adlen and C. Hepburn, "10 carbon capture methods compared", Energy Post, 11 November 2019, https://energypost.eu/10-carbon-capture-methods-compared-costs-scalability-permanence-cleanness/.
15. So far, the irony is that the only semi-economic use at scale appears to be in oil-extraction; see Harrington, "World's largest carbon capture project launches in Texas".
16. Household adoption of solar panel incentives being a case in point.
17. A point Mark Jacobson also makes; see "Stanford study casts doubt on carbon capture", https://news.stanford.edu/2019/10/25/study-casts-doubt-carbon-capture/.
18. See Ritchie & Roser, "Greenhouse gas emissions".
19. "Estimated global anthropogenic methane emissions by source, 2020", https://www.researchgate.net/figure/Estimated-global-anthropogenic-methane-emissions-by-source-2020_fig1_341017757.
20. According to the Energy Transition Commission, the top countries for oil and gas related methane emissions are, by size, Russia, the United States, Iran, Turkmenistan, Iraq, China, Algeria, Libya, Venezuela, Canada, Saudi Arabia and Nigeria. These are the countries whose capacity to regulate and enforce we will be relying on to manage fugitive emissions. See Energy Transitions Commission, "Keeping 1.5°C alive" (September 2021), 32. The one element of fugitive emissions that isn't solved by collapsing fossil fuel demand is emissions from closed coal mines, which continue to emit significant amounts of methane unless flooded.
21. M. Poinski, "Higher plant-based milk prices are justified, but dairy milk is too cheap", Food Dive, 12 February 2021, https://www.fooddive.com/news/study-higher-plant-based-milk-prices-are-justified-but-dairy-milk-is-too/594744/.

CONCLUSION: SPEAKING WITH ONE VOICE

1. C. Munger, "The psychology of human misjudgment", https://jamesclear.com/great-speeches/psychology-of-human-misjudgment-by-charlie-munger.

Index